BEYOND
DOUBT

BEYOND DOUBT

RONALD L. MCCARTNEY, M.S., M.D.

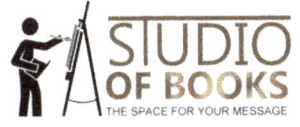
STUDIO
OF BOOKS
THE SPACE FOR YOUR MESSAGE

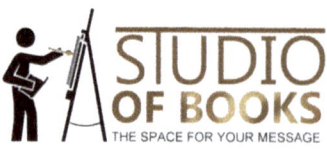

Studio of Books LLC
5900 Balcones Drive Suite 100
Austin, Texas 78731
www.studioofbooks.org
Hotline: (254) 800-1183

Ordering Information:
Special discounts are available on quantity purchases by corporations, associations, and others. For details, contact the publisher at the address above.

Printed in the United States of America.

ISBN-13: Softcover 978-1-964148-56-4
                 eBook   978-1-964148-57-1

Library of Congress Control Number: 2024910878

Table of Contents

ACKNOWLEDGEMENT

To my wife, Donna Dee McCartney for her patient, constructive critique and suggestions.

DEDICATION

The author dedicates this collection of essays (chapters) to the generation of students, educated in our public schools, who have been indoctrinated in a Godless materialistic philosophy, which replaced the traditional moral values previously taught by the morning exercises: reading from the Bible, the Lord's Prayer and saluting the flag, eliminated from public schools in the 1960's. In place of that "religious" framework came the indoctrination of a new secular model of life, said to be "science". In reality it was/is covert atheism, a secular religion in which the autonomous individual, or the state, is god. One of the first results of this change was COLUMBINE, followed by hundreds of other school massacres right up to UVALDE.

This change of traditional moral values alleged that religion is an imaginary, fictional view of the universe, while "science" is said to be reality. The author will show that teaching to be a false dichotomy. True Biblical religious faith is based on evidence, just like true science. Thomas Aquinas had it right.

PREFACE

In this modern era of incredible technical invention and stunning scientific achievement, we are none the less, witnessing historic confusion about our own identity. Human beings have been "male" and "female" even before Moses wrote the book of Genesis, about three and a half millenniums ago. In the recent decade, the facts of biology and history have been challenged, without evidence from either science or history. One might realistically ask, if we can't embrace the reality of male and female, how can we hope to answer the larger query of who we humans are? Where did we come from? Where are we going? Why do we die or do we die?

Pursuit of answers to those seminal questions leads us to eventually embrace, in the big picture, one of two choices (1) we just happened or (2) we were designed and created, male or female!

The author was forced to consider the death question when he was 4 years old, with the death of his beloved puppy and his great-grandmother, in the space of a few months of each other. Well-meaning folks told him that the God had taken his puppy and great grandmother to heaven, (where ever that might be) leaving him to wonder why the Creator didn't just make Himself a puppy and a great grandmother and leave his loved ones with him.

The answers he got to those questions were unsatisfactory, but they started the young mind on a query to find satisfying answers. Eventually he discovered that the knowledge of reality is available and satisfying to the persistent seeker. The author shares his journey to that end in the following chapters.

BASIC UNIVERSAL PREMISES LAW DISCOVERED

PART I

Chapter 1

Introduction to God

My parents lived with my maternal grandparents, for a few years after they were married. The Great Depression was in full swing in 1933 and work was hard to find. Hard times persisted until the Second World War involved the United States. Folks did what they had to do to survive and Dad did what he could to support his wife and four year old son. Pearl Harbor changed our economy by making work available to everybody.

We lived in one side of the old farm house for my first 4 years. There were two kitchens. The smaller one had served as part of an apartment for newly wed family members, until they could afford to find a place of their own. My parents were not the first to have had that arrangement. Mother had ten living brothers and sisters; several of mother's older siblings had used the apartment temporarily.

The old house stood on a level spot about half way up a western Pennsylvania "mountain". Uphill from the house there was an orchard where a variety of fruit trees grew. My favorite tree was the one that had ripe fruit at the moment, but I liked the sweet cherry and the Bartlett pear the best. When they were not available, a red- cheeked apple would do.

Before the garden was planted, I liked to play in the soft tilled soil of the garden plot making roads for my little red car and using sticks for telephone poles along the road-ways. On this particular summer morning I had been alternating road building with riding my trike.

I was home with Dad, while Aunt Dot and cousin's Bill and Dotty took Mother to join their church. Dad wasn't much on going to church, not opposed, just not interested. His mother, my grandma McCartney, was a faithful Presbyterian and Dad would sometimes drive her to church, and leave her there for later pickup.

About the time my stomach began to talk to me about lunch, Aunt Dot, Bill, Dotty and Mother drove up. They were all dresses up for church, like people did in those days. While hugs and kisses were passed around, I noticed that cousin Bill had put my trike up on the porch. Then, the cousins began to explain that Mother had joined their church, and in doing so had become a Sabbath-keeper. So, there were going to be some changes. For starters, I would not be riding my trike on Sabbath; I wouldn't be playing cars and road building either! Those were my very favorite things to do. My world had just disappeared!

Between tears I demanded to know what this was all about. I assume, from my current vantage point that the cousins had been designated to break the news to me, perhaps hoping that might make it more palatable. It didn't. They told me that Jesus was coming soon and that He didn't want little boys riding their trike or playing cars on Sabbath. They said that He was coming back very soon and that if He found me doing those things, I would be burned up instead of going to heaven.

I don't remember hearing anything about Jesus or God prior to this particular day, but I still remember the nauseating combination of emotions that wiped out my interest in lunch and brought a rush of tears and protest, on what had been a beautiful cloudless day. The cajoling of previously well-loved cousins did nothing to assuage all the indignant anger this four year old could muster.

My Grandad Gittins, a tall well-built man, was part-time farmer. He loved the outdoors and hunting. At Thanksgiving time the family, aunts, uncles, and cousins, would gather at the homestead. There would be food without end. Rabbits would be hung from the porch rafters and the cousins would be begging their dads for a rabbit's foot or tail.

Living in Grandad's house, I had noticed that Grandad kept his double-barrel shot gun behind his bedroom door. I had been well cautioned about the power of said gun, and had seen the evidence in the rabbits hanging along the porch roof. In the interest of safety, I had been strictly forbidden to touch the gun.

As the tears subsided, my anger reached its peak. I informed my previously favorite cousins that I wasn't buying any of this new program. I said that I would talk to Grandad about it because I knew that if anybody, including Jesus, tried to harm me, Grandad and the 12 gauge would protect me. Jesus was not off to a good start with me.

Chapter 2

Strike Two

By the ripe old age of 5, we had moved from the maternal grandparents' apartment to a duplex opposite my paternal grandparents, located in a small town about three miles away. My grandmother McCartney took care of her mother, great grandmother Jackson, who lived with her in the duplex. Grandfather McCartney was away much of the time with his work drilling gas and water wells, sometimes as far away as Kentucky.

The duplex was located on a very busy two lane highway which carried all the truck traffic from the coal mines, which were located about 10 miles away in the little town of Clinton, to the steel mill and electric generation plant on the Ohio River. In the daylight hours, the road was busy with heavy truck traffic and our front door was only about 30 yards from the highway. The location gave my parents an excellent reason to resist my incessant request for a puppy. However, to a 5 year old, the space between our front porch and the highway seemed plenty of space for a puppy to play.

Adults are, however, subject to "request fatigue" sometimes known as pestering. This phenomenon is largely due to the fact that parents have lots of important things to do and kids have lots of time to think about what parents should do for them. Additionally, kids seem to be born with a 6th sense of how to wear down parental resistance. One of the most effective techniques children master is the enlistment of outside help. For example, Billy and Jilly have a puppy, why can't I have one?

The lovely lady next door bred Cocker Spaniel puppies and she just happened to have two unsold copies. She got $50 dollar apiece for them. That price might as well been a thousand. She told my parents that she thought every boy should have a dog (I loved her!), but we didn't have the money for one of her dogs.

Having moved to the duplex several months after the Sabbath incident at the farm, cousin Dotty had been reinstated as my favorite cousin, and she came to visit us in our new digs. In a confidential conversation, she told me that one of her neighbors had several puppies, of unknown heritage. She was sure they would be willing to part with one at the right price-free. My dad believed that thoroughbred dogs were usually weaker than mongrels, and he had used this belief (and their price), as an argument against the Cocker Spaniels next door. The combination of puppy linage, price, and Aunt Dot's promise of free delivery, with the prospect of peace in his lifetime, did the trick. I got my puppy!

The consensus of the non-experts I knew, held that the puppy was a mixture of Wirehaired-Terrier and Mexican fence jumper. I could have cared less about the canine pedigree. I loved that puppy and he seemed to love me, especially at feeding time. He ate everything I could smuggle to him. He slept in a box at my bedside, in reach of my hand. He chased me around the back yard and played tug-of- war with an old towel. Like a good dog daddy, I made sure he had food and water and that he didn't "go" in the house.

Paradise lasted a few weeks. The routine was to get up early in the morning and take Bootsie out to do his "business" in the back yard. On the fateful last day, instead of taking him to the back yard, away from the busy road, I took him out to the front yard. I had been training him to come when I called his name and my 5 year old wisdom convinced me that I could watch him from the front window and call him if necessary. So what could go wrong?

Enter the neighborhood dogs having their morning romp just across the busy street. Bootsie joined them in a flash. They all seemed delighted to see him, tails wagging as they checked addresses and the like. Unfortunately, his hearing aids having apparently failed, he didn't come when I called. As the meeting broke up, my frantic calls finally got through. But, the coal trucks were coming. So was disaster.

As I watched from the window, the front wheels of the truck missed Bootsie, but the back wheels did not. I heard a thud and his final little "yip". Bootsie had become a flat piece of fur on the highway. The truck driver braked to a stop. It took about a block to stop the loaded coal truck. He came to the front door where mother met him. He was apologetic, assuring Mother that he couldn't avoid hitting the dog, and offered to take the remains away. She said he didn't need to apologize and she thanked him for stopping. The truck driver picked up my squashed puppy, threw him on top of the coal and drove off.

I didn't cry right away. I had trouble breathing. I couldn't eat breakfast, lunch or supper. I was sick to my stomach all that day and most of the following week. I laid on the couch for days, staring at the patterns on the yellow wall paper. I came to hate the color yellow.

Neighborhood buddies came to play but I couldn't. Mother was worried that I couldn't eat and talked to Cleona, who raised the Spaniels next door. Cleona had just gotten a new-fangled blender and she teased me with a chocolate milk shake. It was my first food in several days.

Cousin Dotty came to visit the next week-end. She did her best to comfort me by saying that Bootsie had gone to heaven to be with God. She said that everything that happens is God's will and we must be happy about that. I wasn't. Having been told that God had made everything I couldn't understand why He had to have my dog. Why not make one of His own, I asked. Mother said we don't understand some things so I shouldn't ask too many questions. I didn't find any comfort in her approach and the questions kept coming. God had struck out again! Much later, reflecting on the tragedy, I would come to realize that my reaction was, at least in part, due to my guilt for disobeying my parents and my realization that I had caused Bootsie's death.

My recovery from Bootsie's demise was interrupted by Great-Grandmother's fall down the back porch steps. She was on her morning excursion to the outhouse powder room when she tripped and fell. Some people thought she tripped on her long skirt. That was probably a good guess, but no one saw the fall. Grandmother McCartney heard her cry and called a neighbor. They helped carry her up the stairs and into bed. Mother made me stay inside so that I wouldn't get in the way, and ask more questions she couldn't answer.

Dr. Pearce, our family Doctor, came to examine great grandma and found that her right foot was turned out and painful to move. Doctor Pierce said that meant she had broken her hip and couldn't walk. She would have to be confined to bed. He said that there was nothing to do but to keep her comfortable. My parents knew that some older ladies who broke a hip were confined to bed and died of pneumonia in a week of two. That sequence held in this case too.

The weeks following the fall there was a lot of quiet adult talk, with relatives and neighbors. Most of it was out of the range of my hearing, undoubtedly on purpose, to shield me from things I couldn't understand. The talk would have been about Great Grandmother's imminent demise, where she would be "laid out" and buried. In those days, the deceased were visited by family and friends in their home, lying in their casket or on a couch or bed. Funeral homes were uncommon in little country towns.

In our town the undertaker would take the body to his establishment to be embalmed. Then, the deceased would be taken to the home for visitation, prior to taking the body to the grave yard for graveside services and burial. In consideration for my young age, the visitation occurred at a relative's house up the street; I was not allowed to see her.

Two deaths in about two months! What could God possibly want with Great Grandma? She just sat in her rocker, never did much of anything except peal apples and potatoes. So what that people said she was old? Why do people get old and have to die? What did my puppy and Great Grandma do to deserve to die?

And just what is this "sin" thing that people talk about? Cousin Dotty said it's doing bad things. But I did what I was told, usually, never hurt anybody, loved dogs and ate my veggies. Yes, I did say some of those bad words that adults say, and tell little boys and girls not to repeat. But, I never, ever, used them when around adults.

From the same window where I had called Bootsie to his death a few weeks before, I watched the funeral procession take Great Grandmother to her grave in a big black hearse. The old sick feeling in the stomach came back. Death had made its unwelcome presence known in my young life again. I had bad dreams about the big black hearse for a long time. Things had changed forever. The ultimate reality was very real.

Chapter 3

Introduction to Darwin

Getting satisfying answers to one's questions is not always easy and sometimes it seems impossible. Further, the dilemma resulting from getting diametrically opposed answers from alleged experts can be the ultimate frustration. However, when the frustration leads to further study, the reward can be in finding gratifying answers.

During the interval between losing Bootsie and the seventh grade, I attended weekend Bible school with Mother. I learned memory verses and heard many of the stories appropriate for young minds, derived from the Old and New Testaments. The summation of that teaching seemed to be that breaking God's rules was sin. The religious cause and effect equation was based on the scripture that says that death is caused by sin.

Enter Mr. Leiberman, the seventh grade science teacher with the "truth" about the world and our universe. As you entered his home room, one could see a series of attractive color pictures next to the celling on the left, front and right sides. The pictures were in a sequence that started from the back left. The first picture was of a rain storm with lightening flashing over a swamp. The pictures continued with one-celled animals swimming in the swamp, followed by a series of more and more complicated life forms, which eventually, crawled up on the land. The last of the picture series in the right back corner of the room was a human infant.

Mr. Leiberman directed our attention to the picture progression that everybody had already seen. He said that we would be studying science which he defined as "the way things really are" having been discovered by the smartest people in the world. Truth and reality are discovered by scientists, he said, and we **should** accept what they tell us. He explained that our class would meet five days a week and during that time we would discover some things that might contradict ideas that we had been taught. He said that some parents might be upset by some of the things we learned in class, so there would be no homework and our textbooks were to stay in our desks. He said it would be fine if we did **"weekend stuff"** (church) on the weekends, but 5 days a week in class, there would be no place for it. We were going to be "scientists".

As the class proceeded through the year, Mr. Leiberman presented Darwin's evolutionary view, as he understood it. The textbooks he used stayed in class and there was never any homework, as he had promised. We were taught that human beings are the current highest form of random chance selection that has acted over incomprehensive time passed.

Natural selection is god, if you need one. Death is a necessary component of the evolutionary scheme, so that there can be space and nutrients for the succeeding improved evolutionary generations. As to the evidence for this evolutionary indoctrination: well look at the pictures!

The following year, the biology course was taught by Mrs. Crawford, continued the theme begun by Mr. Leiberman. Having had some time to analyze the Darwinian scheme, I asked her how life had begun in the first place, seeing that science demonstrates that life is dependent on and derived from previous life (parents). I could not believe the loud, venomous tirade she launched at me, forbidding me to bring "religious" ideas into her science class. She threatened that if I didn't learn to think "scientifically" I would have to repeat her class and my future in any scientific endeavor would be over before it got started. Needless to say, I never asked another question in her class. At recess, several classmates expressed their condolences and swore off ever asking anything themselves. My friends and I could not remember my asking anything about religion.

This incident advanced my growing conviction that adults didn't have all the answers and some of them highly resented anyone who brought that fact to light. This also led to other questions. Is evolution theory above questioning? If evolution has the answer to important questions like the origin of life, why not just give the answer? If it doesn't have the answer, why all the anger with a student for asking?

Chapter 4

Darwin Close Up

Darwin's study of animal life in the early 1800's led him to document the far ranging capabilities of plant and animal life to accommodate to environmental needs by adaption. He documented his findings, like a good scientist, on his journey along the coasts of South America and in the Galapagos Islands. His observations lead him to consider how far the adaption he witnessed might go, in explaining biologic diversity. From 1835 to 1859 he developed a theory to explain his observations.

His scheme proposed that life began by some unknown means, perhaps by accident, and progressed from simple to complex over time. Complexity, he thought, could be achieved by small, gradual changes in plants and animals, which enabled them to survive changes in their environment. The survivors, he thought, could then pass those adaptive changes to their offspring. The less able to adapt varieties perished, and the best able survived; hence, SURVIVAL OF THE FITEST. Sometime later, with the discovery of genetics and cellular complexity, Darwin's disciples would add the idea that mutations also contributed to the development of increasing complexity. (Neo-Darwinists)

Darwin had reservations about his evolutionary ideas. For starters, many of the scientists of his day believed in the Biblical flood, and the chronology which dated the age of the earth at several thousand years. He believed that his scheme would take much longer than that, since it was not observable in real time. Additionally, his theory contradicted some of the best, time tested truths. In the thirteenth century, Thomas Aquinas, using logical arguments beyond rebuttal, confronted and refuted the academics' assertions that human reason is superior to revealed knowledge. Considering the idea that human reason could arrive at any truth without revelation Aquinas challenged:

> There are two distinct classes of truth, the natural and supernatural. Natural truths are those which can be known by the reason of man unaided by revelation. Man, by the power of his intellect without any especial assistance from God, can know many things about himself and the material world, and about God. Supernatural truths are those mysteries, hidden from ages and generations in God, which searched all things, even the deep things of God. Hence, it is wrong to suppose with Abelard and other [Rationalists] that all things are to be measured by the mind of man. **Our intellect is finite**, but it tells us with certainty that **God is infinite**, hence we could prove a priori a truth which we know from experience, that there are more things in heaven and on earth than are dreamed in or philosophy. D.J. Kennedy: Thomas Aquinas and Medieval Philosophy. 1069-99 (emphasis mine)

In 1802, William Paley published a book called NATURAL THEOLOGY, a treatise introducing the deductive logic underlying what has come to be known as **Intelligent Design.**

> There cannot be design without a designer; contrivance without a contriver; order without choice; arrangement, without any thing capable of arranging; subservience and relation to a purpose, without the which could intend a purpose; means suitable to an end, and executing their office in accomplishing that end, without ever having been contemplated, or the means accommodated to it. William Paley. Natural Theology. 484-86

Paley's book is an expanded commentary on Paul's letter to the Romans.

> For since the creation of the world His invisible attributes are clearly seen, being understood by the things that are made, even His eternal power and Godhead, Rom.1:20 NKJV. (It is thought that Paul was referencing David's Psalm 19, "the heavens declare the glory of God.")

Paley's book quickly became the accepted textbook on applied logic, and was required reading for entry into the universities. Darwin read it. So, how is it that Darwin's Atheistic Materialism ideas survived and prospered, in spite of Paley's powerfully reasoned treatise and the opposition of the Christian Church? Darwin's idea caught on, in part, because he linked his theory of an unguided spontaneous Epicurean type Naturalist creation, to the well-known practice of selective breeding. Thomas Huxley helped by promoting Darwin's idea as "science" to the London scientific elites.

Darwin asserted that Natural Selection could bridge the barrier that separated species, if given enough time, in spite of the fact that he knew that mules and ligers are sterile! His friend Lyle convinced him that The Genesis record was unreliable and that Hutton's uniformitarian idea would give enough time, (millions of years) for natural selection to produce the undemonstrated evolutionary changes that Darwin proposed. Substituting the term "natural selection" for the similar Epicurean "concatenation", gave his idea a scientific ring.

Darwin worked on his evolutionary idea from about 1835 to 1859. He hurried it to publication in 1959 to avoid being scooped by a friend and competitive contemporary, Henry Wallace. He had chosen impersonal, undirected chance natural selection (evolution) over intelligent design.

Uniformitarianism seemed to solve one of the several problems bothering Darwin, by alleging our earth to be millions of years old. He still had no explanation how life started. Additionally, he needed some evidence from the fossil record to bolster his idea about undirected development of mankind from single cell animals.

He expected fossil evidence from the large number of transitional animals that his theory proposed. Those fossils have never been found, nor has the evolutionary answer to the origin of life.

In spite of sophisticated laboratory experiments, all attempts to create life from the life-less elements, including laboratory attempts guided by intelligent humans, have failed. So, how might life happen by unintelligent chance?

Darwin advanced his theory in spite of the unsolved problems, hoping that future study of the fossil record would document the vast numbers of fossils that could prove the existence of the intermediate life forms needed to make his theory credible. On the pretext of ADAPTABILITY of the flora and fauna, Darwin advanced his theory of evolution by chance "natural selection". In doing so he embraced agenda science (scientism), the religion of atheists.

Over one hundred years from the publication of Darwin's ORIGIN, and in spite of extensive excavations in fossil beds around the world, not one of the necessary transitional forms have been found. Read none. There have been many fossil frauds introduced by Darwinian evangelists, like the infamous Piltdown man, Nebraska Man, Java man, Lucy and more. About a hundred years of searching without success led noted evolutionist Stephen J. Gould to put the search for intermediate fossils to rest.

There are no transitional forms, he admitted, so stop looking. He then proceeded to destroy the Hutton Uniformitarian Theory by proposing an explanation for the absence of transition forms. He fantasized that biological evolution takes place episodically, and at such a rapid pace that it is not observable, and it therefore leaves no fossil remains! That sounds catastrophic to me. He called the idea "Punctuated Equilibrium". The unvarnished truth is there is no evidence of so called "intermediate fossils." The earth's geological history and the fossil record validate Noah's world-wide flood.

Darwin thought that the cell was the smallest unit of life, a kind of amorphous gelatinous glob. He thought that it could be morphed into something else by its environment. Knowledge of the intricate, intra-dependent protein factories of the cell had to await the electron microscope, the DNA code and other scientific tools.

Extensive experimentation with induced mutations began within a few years of Roentgen's discovery of X-rays in 1895. Their mutagenic capabilities were discovered soon after in the early twentieth century. Extensive experimentation with fruit flies in the early twentieth century showed that nearly 90% of induced mutations produced undesirable or fatal outcomes. The studies showed that genetic hybridization and induced mutation reduces the amount of desirable biological information. Therefore, it cannot add to or gain either the quality or quantity of information necessary for Darwinian evolution to occur.

Twentieth century scientists looking deeper into the cell discovered the complexity of the cell membrane and the internal organelles. The cell's protein factories show astounding interaction and relationships so complicated that random chance development is impossible. Then, it was discovered that the DNA code control of cell function and reproduction is degraded by hybridization and mutation. There is no demonstrated method whereby sufficient new information can be added, to change the basic structure of a cell, by the proposed evolutionary means. The observed facts of cell biology support the law of entropy. Devolution is the law of our natural world. Evolution as a natural creative force does not exist. It is only the inventions of man that demonstrate the evolutionary process of simple to complex.

So, the impersonal and mechanistic Darwinian answers to our questions about life and death leave us with the following conclusions; we don't know where life came from or where it goes. Death is a necessity for evolution to do its alleged work. The purported new life forms being created need the food and the space of the forms being replaced.

Darwin's evolution appears to be revised Epicureanism, "[E]at, drink and be merry for tomorrow we die". (English idiom attributed to the Greek philosopher Epicurus)

Mr. Leiberman and Mrs. Crawford defined science as if it were a synonym for reality, the truth. But their truth, although advanced with evangelistic zeal, openly challenged the ideas of people (family) that I trusted, and even when I didn't like what I heard from my family, I still trusted them. By disrespecting the people I trusted, and belittling their "weekend stuff", Darwin and his disciples lost my confidence. I learned that there are usually multiple answers to any question that one might ask, and sometimes more than one of them might prove satisfactory. But, when two answers are exact opposite, one of them is most likely wrong. I decided to keep looking to find out which one of the contradictory ideas was correct. Was it mechanistic impersonal chance (natural selection) or intelligent design?

Chapter 5

Univeral Law-Physics 101

The mail train, the Midnight-Flyer, went through the Fifth Avenue rail crossing in Coraopolis, at about mid-night. The train, it was estimated, was traveling about 80 mph when it hit Harvey's car that Saturday night in March, a few months before his graduation from high school. Harvey was taking his fiancée home from a dance. The car was thrown about 150 yards. The closed casket funeral for the couple was private.

Harvey was more mature than most of his classmates in the graduating class of 1951. As a junior, Harvey rode the same school bus that I did. His stop, at Stoops Ferry Station, was about half way from my home to the Moon Township High School. Harvey had a cute little girl friend who lived in the same housing development and they always sat together on the bus. Since she was younger, a sophomore, Harvey got teased about robbing the cradle.

Harvey had a summer job in a local department store that his father managed. During the summer between his junior and senior years, he acquired a late model used car. When school started his senior year, he drove his car to school and his girlfriend sported an engagement ring. They were to be married when Harvey graduated in June. It was rumored that they had already found a house to buy.

I don't remember voting on which of our classmates was most likely to succeed, but if we had voted, Harvey would have gotten most of the votes. In many ways that seem important to high school students, he had already succeeded. Just count the ways: nice car, nice girl, a good job with dad, a house. What more is there?

Monday morning after the accident was a very solemn time at the Moon Township High School. Harvey's home room was quiet as the students passed around copies of the local newspaper and tried to absorb the details of the tragedy. Speculations on how this could have happened at this familiar crossing ran the gamut, was the barrier down, did he go around it, did the lights work, did he know when the train came through town? No end to the questions and no answers to any of them.

Harvey and I had both had physics class as juniors. We had learned some of the fundamental laws of nature like; two material objects cannot occupy the same space at the same time without one or both of them undergoing deformity. Additionally, there is the equation about force being the product of speed and mass. Did Harvey know these laws? Did he believe them to be true? Did he knowingly disregard them or was he distracted? Was there fog? Did the train sound its horn? Did any of that even matter now? Harvey broke natural law and paid the price. His seat was empty as the graduating class took theirs.

There is a price to be paid for breaking natural law. The most basic principle of universal physical law is called, "cause and effect". Everything we observe has a causation, to every action there is a reaction, nothing just happens.

Another occasion for questions about God, law, life and death. More questions to ask and answers to find.

Chapter 6

The Law of Being

I was about 12 years old when the obsession to drive a car began to consume me. Since age 7 or 8 years old, I would spend time on weekends or evenings sitting in the driver's seat of the family car, a 1938 Chevy, fantasizing about driving to places near and far. I was allowed to shift the floor mounted gears, provided there was no resistance. It was enough until I was about 12, when my friends started talking about their driving experiences. I entered the 11th grade at age 15 and since my birthday was in October, I could get my license early in the school year.

Uncles Frank and George had let me drive their cars when I was 15. Uncle Frank let me drive his '39 Ford in a level field near where he worked; I walked on air for about a week afterward. Uncle George had me drive his '35 Ford to pull a scoop he used to dig the foundation for his new house. I was ready!

Our '38 Chevy was well used and needed frequent attention. Dad was a good mechanic, so I was able to watch and occasional help him keep it running. He rebuild the engine, replacing piston rings, rod bearings, valves, spark plugs and breaker points. When the howling noise in the differential became intolerable, he replaced the pinion bearings.

After rebuilding the Chevy, repairing the house, and paying off the mortgage, Dad decided it was time for a new car. It would be his first and just in time for me drive it, since I had just become a licensed driver. What timing. What a great start to my junior year in high school.

The new car was a step up in the world of cars. Pontiac introduced their new "Silver Streak" models, 6 and 8 cylinder models in 1950. They were available in two tone paint with two and four door sedans. My parents chose a medium blue, two door six cylinder sedan. It was quiet, rode and drove like a magic carpet. The gear shift leaver was moved from the floor to the steering column. The cherry on the ice cream sundae for me was that I had a newly minted driver's license and got to drive the Pontiac to school one day.

Early in my junior year, shortly after I got my driver's license and after we got our new car, a neighbor who lived about a half mile down the road, got a new two tone green, straight eight, silver streak, hydramatic four door Pontiac sedan. It had a radio! A big deal in those days. So, my school mate got the top of the line model, we had the bottom. But we loved our new car, it was a giant step up for us. The interesting thing was that this neighbor was a kid like me, maybe a year or two older, but in my grade, and the car was his! He lived with his mother, in what everybody called the "log cabin".

Nobody knew much about the family situation. Only the mother and son lived in the cabin. There was a barn near the house that was twice as big as the cabin, occupied by some chickens and a few pigs. There was no grass around the cabin but seasonally there was a large garden between the cabin and the road. Spring and summer, a woman in a dark long sleeved, long skirted dress could be seen working the garden or selling some produce from a roadside stand. She wore a babushka;

unique in our community. The bright shiny new Pontiac looked entirely out of place in its parking space by the barn. The question was, "Where did this high school student get the money to buy a top of the line Silver Streak Eight?" Nobody that we knew had the answer.

One morning, after getting my driver's license and our new car, the school bus stopped at the cabin as usual. The two tone green, straight eight Silver Streak hydramantic was seen parked in the driveway to the cabin. The young student, whose name I never knew, was sitting in the driver's seat of the Pontiac, instead of standing in line to get on the bus. When the bus started out, he roared out of his driveway, throwing dust and stones, as he gunned the straight eight around and in front of the bus. Driving ahead of the bus to the next stop, he parked there while two of the waiting students got in. When the remaining students got in the bus, he launched again, in a cloud of rocks, dust, and squealing tires. He continued that routine for some weeks, then we would only see the Pontiac parked at school. He seemed to be living the fantasy life every teenager would covet.

High school senior at last! On the bus to school, the first stop was the cabin. Every eye on the bus opened wide in amazement at the sight of the two tone green, straight eight, hydramatic, four door Silver Streak Pontiac, parked on blocks by the barn. The tires were flat, the hubcaps gone, hood up and car covered in mud. A total wreck in less than a year. No one that I knew was close enough to the student to know what happened. The student had apparently dropped out of school. He was not in my senior classes and did not graduate with our class.

In the months that followed the death of the Silver Streak Pontiac, some pieces of the puzzle came to light, reportedly from the driver of the tow truck that dragged the dead Pontiac from the scene of its demise, to its ignominious resting place with the chickens and pigs. The story was that the straight eight had stopped suddenly during a drag race, the engine smoking and the car motionless. It could not be started. There was a very bad smell from under the hood and when they finally opened the hood, smoke and steam belched forth.

After the engine cooled, the oil dip stick was found to be welded in place by the intense heat of a burned up engine. Someone who knew the driver, said that he had not changed the break-in oil at 500 miles as required. All new cars, at that time, came with the requirement. Furthermore, he apparently never checked the oil or coolant levels. He put gas in the tank and drove like a madman, challenging everyone he could to a drag race. He never washed the car, checked the tires or performed any maintenance as proscribe in the owners' manual, if he ever even read it. In sum he violated the natural law of the machine, as written by the designers, engineers and manufacturer of the Pontiac. He killed the dream machine.

Reflection on the circumstances surrounding the death of the Silver Streak Eight has lead me to the observation that the creation of any material object results in a body of specifications which define the created object, and in fact constitute its "law of being".

Another example would be that of a potter who forms a ceramic vessel from an amorphous lump of clay, creating its shape, glazing, and firing it. If the vessel is subsequently dropped or otherwise broken, it is destroyed. If the potter had dropped the blob of clay before he had formed the vessel, no harm would have been done to the clay. Creating the ceramic vessel, imposed restrictions on it that constitute the "law of being" for that creation.

Dad took care of his Pontiac as prescribed in the owner's manual that came with the car. The break-in oil was changed at 500 miles and the subsequent oil changes met the recommended weight, grade and interval specified in the Pontiac "Bible". The car served well for many years, as intended.

The death of the Straight Eight made me consider the possibility that we humans are subject to our own "law of being" like everything else we observe in the material world. If we find evidence that we were created, wouldn't it be scientific to assume that we were made, like everything we observe in nature, with abilities and limitations that would be the result of our "law of being"?

Chapter 7

Looking for Answers

My chosen college was about 250 miles away. That was farther from my home that I had ever been. I had chosen the pre-med course of study, inspired by my dad and our family physician, Dr. Pearce. My world was changing in a big way and faster than I wanted. The friend who had accompanied me from home was to be my roommate. We went to register for classes that first morning, pre-med for me and nursing for him. We agreed to meet at the room after registering, so we could have lunch together. When I got to the room, he was nowhere to be found. His things were gone too, and the monitor said that he had called a cab about mid-morning. He was already home-sick, and took off to the bus station and home. I was home-sick too, but I was motivated to find some answers and do some things that couldn't be done at home.

Chemistry, physics, math, biology and history are studies that operate within the bounds of observable natural law. The learning conducted within these bounds gives us trustworthy information and an appreciation for the world and the universe we observe. The reliability of the things we learn in these disciplines is proportional to the quality and quantity of the information on which they are based. Science, I learned, is about finding nature's laws through observation, constructing a theory to explain the observations and then testing the potential law to either prove or disprove it.

Johannes Kepler, 17th century mathematician and astronomer, characterized science as "thinking God's thoughts after Him". That, of course, is a tall mountain to climb, given the great chasm between the capacity of God's mind and that of mankind. (Quote Fancy.com)

So the proper role of science is to seek the answer to questions about the universe that can be answered by repeated human observation and testing. Questions about origins, however, must be answered by disciplines other than the "scientific method" because the creation was a one-time event with no human being present to observe it, or record how it happened. Since science can only be credible with matters observed in time and space, we need to know something about history, religion and philosophy in order to establish an intellectually defensible world view and universal perspective.

Religion can be defined, "[A] cause, principle or system of beliefs held to with ardor and faith". (Webster) Religion deals with matters involving the transcendent which are outside the capability of discover by scientific methods. Science deals with the material and temporal whereas religion deals with the unseen and reality beyond time and space. Nearly all human societies across the span of recorded time have had some form of religion. The single most important reason for the ubiquity of religious belief and practice is the universal presence of death. Death is the ultimate datum.

The comparative study of religions, both ancient and modern, confirms the pre-occupation of thinking humans with the origin, purpose and destination of humanity. For example, the great Egyptian civilization has left us enormous pyramidal structures that testify to their preoccupation with the after-life. Their extensive preparation for the "underworld" consumed much of their earthly time and treasure. The presence of similar structures, especially in the middle latitudes around the globe, indicate the universality of the Egyptians preoccupation with death. Two events universally witnessed (science) are birth and death.

The first is accompanied by wonder and awe. It usually brings joy and anticipates good things to come. The second is also accompanied by wonder, but is usually followed by boundless grief and unspeakable loneliness.

Life and death, mysteries for which science has no answer. The answer to some of life's most pressing questions can only be found in the world of the transcendent and the metaphysical. As far back as recorded history can take us, men have sought escape from the inevitable. The most common solace, embraced by nearly all humans, has been the idea the death is not the end of being. Both ancient and primitive societies have developed funeral and burial practices to supposedly prepare the deceased for the future life. Witness the tombs, burial mounds, caves and pyramids which include material objects to be used in the "next world".

Reason and logic are the tools of honest intellectual exercise in both the temporal and eternal areas of life. When they are used in conjunction with cause and effect law, the result is good science and good religion. Comprehensive reality depends on both credible science and sound religion. Scholarly evaluation of history, religion and tradition can result in the discovery of reality as surely as the most elegant scientific experiment.

I hoped to find if religion could provide satisfactory answers to the big questions, and to that end I scheduled all my elective hours in classes that taught comparative religions, Bible, and Western denominational Christianity.

By graduation time I had acquired some answers and a lot more questions. Like, why are there so many Christian denominations, all of which use the same text, the Bible? Then, one can find contradictory doctrines and all shades of interpretation on almost any topic. I remembered that my Dad had had problems with the diversity of religious opinion, theoretically based on the same text.

Nonetheless, a comparison of the documents on which the world's non-Christian religions are based, shows them not comparable to the Judeo-Christian "Bible". The Judeo-Christian scriptures have withstood the intense criticism of doubters for centuries and proven to be accurate, reliable, historical and inspirational.

I concluded that religion and science both have Achilles heels. Darwinists are the example in science; they use whatever science supports their **agenda**, and they wave the science flag over their doctrine, i.e. mechanistic, impersonal change over time. Some religionists search to find support for the things they wish to believe, and they wave the religion flag over their **agenda** doctrines. That behavior leaves the serious student to search for the evidence from which to build a satisfying, intellectually defensible reality in both science and religion.

When a person whose profession is science or religion refuses to admit the need for falsification for any theory or doctrine they advance, they have abandoned scientific methodology and/or sound exegesis. Their conclusions need to be examined in the light of the available facts. Everyone has the God given right to their own opinion, but to advance ones unproven and/or unprovable ideas authoritatively, without objective facts and/or evidence, is a form of intellectual fraud.

Chapter 8

Prescience, the Absolute Proof

In the fall of 1959, it was about 11 O'clock at night, when the OBGYN resident at the Los Angeles County Hospital in L.A. California, told the Midwife on duty that he was going to bed. The usual orders were in place; the medical student (me) on duty would take first call on all uncomplicated deliveries.

Shortly after falling asleep, The phone rang, and the Midwife said she had a patient fully dilated and in active labor. By the time I gowned and gloved, there was a little tuft of black hair in sight. With thumb and index finger securely around the baby's neck, the delivery was safely completed and a cursory exam showed a normal male. As I turned back to deliver the placenta, the Midwife exclaimed, "There's another one!" She was right; a second, unanticipated little boy soon joined his brother. When I congratulated the mother on her twin boys, she asked if she would get "green stamps" too. She already had two children at home and was less than thrilled with the sudden doubling of her family. This was before ultrasound, which has largely eliminated such surprises.

My first delivery of twins was my last delivery at the L.A. County Hospital, and made it a memorable night. I had held newborns in my hands 53 times during my rotation and by that experience got to observe the absolute proof that Darwin was wrong. The human birth phenomenon demonstrates **prescience; adaption** cannot explain it. The plan for the creation of life that we observe in the birth phenomenon (science), requires a level of intelligent design and foreknowledge that is clearly demonstrated by the emergence of a new life. Paley used the word "contrivance" to describe the process of arranging for future need.

> I can hardly imagine to myself a more distinguishing mark, and consequently, a more certain proof of design, than preparation, i.e. the providing of things beforehand, which are not to be used until a considerable time afterwards; for this implies a contemplation of the future, **which belongs only to intelligence.** (emphasis mine). William Paley, Natural Theology, 484-6.

Adaption to an environmental need may be explained as "survival of the fittest", but provision for an anticipated future need is evidence for an intelligent design that is the signature of the Life Giving Creator. Such prescience is worlds beyond the capability of any proposed adaptive evolutionary scheme.

When the origin of our human existence is discussed and all the various theories about how we came to be here are heard, there are only two possibilities: (1) An Outside our universe designer/creator did it, or (2) impersonal forces inherent to "nature" originated everything we see.

A study of comparative religions makes it clear that the Bible account of origins is unique. The Scriptures introduce the "outside our universe Creator" concept that stands in direct opposition to the Greek and Darwinian materialistic assumptions. Darwin's theory of Evolution is built on the Greek assumptions and is expressed as three propositions: (1) That the earth's history is uniform, giving unlimited time for changes in nature that no human can observe, (2) That progressive development and complexity of flora and fauna are the result of **adaption** by impersonal environmental forces and (3) That the species with the **most surviving offspring** survive and thrive so that species can "evolve".

The study of our earth's history, past and present, shows it to be CATISTROPHIC, in direct contradiction to the Uniformitarian Doctrine on which Evolution depends. I wonder if Darwin ever read about the volcanic eruption at Pompeii? We have Krakatoa and Mt. St. Helens for additional examples in our lifetime.

Darwin's second assumption is as problematic as the first. As a child develops and is about to be born, it is dependent on the mother for oxygen and nutrition provided by the placenta. The source of oxygenation changes instantly from the mother to the child, within seconds of birth. The baby exits the womb (unclothed), evicted from its 98.6 water bath to a dry temperature some 20 degrees cooler. A gasp for air (or protest) instantly activates the previously unused lungs, according to the genetic PRESCIENCE. For about 9 months, the child has been developing from a fertilized ovum to an infant human being, during that development, totally dependent on the mother for food and oxygen. The plan is for the infant to eventually live independent of the mother. So the plan develops lungs and internal vasculature that are UNNEEDED while in the mother. That process is not driven by **evolutionary adaption** or need. Planning for the future (design) is the mark of INTELLEGENCE at work. Evolutionist insist that their *explanation* of increasing complexity of organisms occurs without intelligence. *I agree with that!*

Upon the first cry, the infant's lungs inflate. The Foramen Ovale shunt in the right atrium closes, bringing blood to the lungs for oxygenation. The Ductus Arteriosis that shunted blood from the pulmonary artery to the aorta closes. All according to PLAN, in a matter of seconds. Similar *anticipatory*, genetically programed provisions are made for the teeth, the eyes and the endocrine feed-back system. There is no time for Darwin's *adaption* mechanism to work! It all has to happen correctly, the first time, according to a plan for **future** need.

For You formed my inward parts;

You covered me in my mother's womb.

I will praise You, for I am fearfully and wonderfully made;

Marvelous are Your works,

And that my soul knows very well.

My frame was not hidden from You,

When I was made in secret,

And skillfully wrought in the lowest parts of the earth.

Your eyes saw my substance, being yet unformed.

And in Your book they all were written,

The days fashioned for me,

When as yet there were none of them. Ps 139:13-16

As to the third premise of the Evolutionary idea, the development of sexual reproduction, contradicts the *"most offspring"* idea because asexual reproduction produces many more offspring than sexual reproduction ever could. Furthermore, the probability of developing male and female animals of the same species at the same time and place in our wide world, and doing it multiple times for the different species, stresses credulity and probability mathematics to their limit.

Dawkins has said that nature looks like it was designed. He is right about that! In spite of his disdain for the scripture, his observation agrees with Moses, David and Paul. He should believe his eyes, that's the mark of a Scientist.

So God created man in His own image; in the image of God He created him; male and female He created them. Then God blessed them, and God said to them, "Be fruitful and multiply; Gen 1:27-28

The heavens declare the glory of God;

And the firmament shows His handiwork.

Day unto day utters speech,

And night unto night reveals knowledge.

There is no speech nor language

Where their voice is not heard.

Their line has gone out through all the earth,

And their words to the end of the world.

In them He has set a tabernacle for the sun,

Which is like a bridegroom coming out of his chamber, Ps: 1-5

For the wrath of God is revealed from heaven against all ungodliness and unrighteousness of men, who suppress the truth in unrighteousness, because what may be known of God is manifest in them, for God has shown it to them. <u>For since the creation of the world His invisible attributes are clearly seen, being understood **by the things that are made**</u>, even His eternal power and Godhead, so that they are without excuse, because, although they knew God, they did not glorify Him as God, nor were thankful, but became futile in their thoughts, and their foolish hearts were darkened. Professing to be wise, they became fools, and changed the glory of the incorruptible God into an image made like corruptible man — and birds and four-footed animals and creeping things. Romans 1:18-23 (emphasis mine)

Reality science, like reality religion is based on EVIDENCE and there is no conflict between the two.

Now **faith** is the **substance** of things hoped for, the **evidence** of things not seen for by it the elders obtained a good testimony. Heb11:1-2 (emphasis mine)

THE GOD OF CREATION SCIENCE

PART II

Chapter 9

Judeo-Christian Theology

We are the beneficiaries of scholars who, over millennia, have studied and preserved the manuscripts of biblical authors. The result of their dedication, hard work and sometimes martyrdom, is the many translations and versions currently extant. In the early A.D.s (100-200), Jewish and Christian scholars selected the manuscripts they judged to be inspired by God, and codified them into the essence of our Bible. Subsequently, men like Martin Luther, Tyndale, Wycliffe, King James and others oversaw the translation from the original or previous languages, to German and English respectively.

There is a reason that the Bible (Judeo-Christian Scriptures) is the most published document in the world. It is by far the most comprehensive treatise on the topics of good and evil. It covers human origin, purpose, destiny, and life comprehensively. It claims to be The Word of the Creator God, written by servant-authors, whom He has inspired to document the relationship He desires with His creation, from the beginning, to the present and into the eternal future.

What we call a "book" (Bible) is really a collection of books, written by three dozen or so authors, whose writings cover about 1500 years of history from 1400 B.C. to 100 A.D. Moses wrote the first five books, starting with GENESIS, the account of the original creation. The Apostle

John wrote the last book, The REVELATION of JESUS CHRIST, where we find a description of the new earth. Moses was educated to be an Egyptian Pharaoh, John was very likely a fisherman. In 66 books, the history of the world is chronicled, by authors of diverse backgrounds, inspired by God, to announce the end of death and the triumph of life in Eden restored.

I do not know Greek, Latin, Hebrew or Aramaic, so I/we are indebted to many great scholars, some of whom gave their lives, to bring us what they called "The Word of God". I/we are also indebted to many recent Bible scholars and commentators. In the chapters that follow, I will share my search for answers to the questions I encountered at a very young age: Will Jesus really burn up little boys who ride their tricycles on Sabbath? Why does He smash puppies and push old ladies down the steps so He can take them for Himself? Childish questions to be sure, but the answers are important to everyone. Here is a big boy question, why did God let Hitler, Stalin and Moa kill about a million innocent civilians in the twentieth century, and how about Vladimir Putin's genocide in Ukraine in our own time.

Any claim made in the name of science needs to be validated in order to be credible. So, it should follow that the "Word of God" should also be validated. Testing of the "word" has been done continuously over the years since the Bible was assembled about 200 A.D. This started with evaluation and selection from the many manuscripts available. Secular history and archeology have been used effectively to refute the false claims of the ever present critics. Several examples of validation will suffice to illustrate the point.

Many ancient cites mentioned in Old Testament scripture do not exist today, or in recent history. That fact has been cited by critics in times past as evidence that the Bible is unreliable. However, most of those cities have been identified by subsequent secular history and or archeology. Examples would include: Tyre, Sidon, Ninevah and Sodom.

In the New Testament, the critics said that the references to a synogogue in Jesus's home town of Nazarus proved the scripture in error because, they declared, there was no synogogue in Nazarus. Archeologists recently dug up its foundation.

The Old Testament book of Daniel contains a detailed prophecy about the destruction of Jerusalem which occurred in A.D. 70. The critics said the prophecy had to have been written about **100 A.D. because it was too specific and accurate** to have been written when Daniel lived several hundred years B.C.. But, the same critics found copies of Daniel's prophecy in the Death Sea Scrolls and dated them at **200 B.C.** These examples are only a small sample of the many available. Apologize anybody?

In order to have confidence in what scripture leads us to believe, it is important to be convinced as to its authority and accuracy. Then, with the multitude of Christian denominations and sects in mind, one might ask, if perhaps, Bible scholars need to avoid the mistake that some material scientists make. They look for evidence to support what they **would like to believe,** instead of waiting to base their beliefs on a comprehensive consideration of the evidence.

One of the problems to avoid in Bible study is jumping to the conclusion that the diverse comments of different authors, on the same subject, constitute a contradiction. In complex, substantive issues, the whole picture may require consideration of the times, places, authors' perspectives, languages, translations education and other circumstances. Since all scripture is the result of God's inspiration, it is our job to construct the correct picture using the author's different perspectives; keeping in mind that the Holy Spirit directed the authors and He cannot be impeached.

All Scripture is given by inspiration of God, and is profitable for doctrine, for reproof, for correction, for instruction in righteousness, that the man of God may be complete, thoroughly equipped for every good work. 2 Tim 3:16.

Jesus used parables for much of His teaching in order to confuse His enemies in the Jewish hierarchy. Sometimes the disciples were also confused.

Then His disciples asked Him, saying, "What does this parable mean?" And He said, "To you it has been given to know the mysteries of the kingdom of God, but to the rest it is given in parables, that 'Seeing they may not see, And hearing they may not understand.' Luke 8:6- 10

It is important to understand that a parable is a fictitious story constructed for the purpose of teaching a specific truth or doctrine. Jesus used that communicative tool to avoid confrontation with His enemies and to help the listeners remember the lessons. Critics, looking for "errors", have sited the following parable as proof that Jesus didn't know anything about agriculture, so He couldn't be the Creator of it!

The Parable of the Wheat and the Tares

Another parable He put forth to them, saying: "The kingdom of heaven is like a man who sowed good seed in his field; but while men slept, his enemy came and sowed tares among the wheat and went his way. But when the grain had sprouted and produced a crop, then the tares also appeared. So the servants of the owner came and said to him, 'Sir, did you not sow good seed in your field? How then does it have tares?' He said to them, 'An enemy has done this.' The servants said to him, 'Do you want us then to go and gather them up?' But he said, 'No, lest while you gather up the tares you also uproot the wheat with them.

Let both grow together until the harvest, and at the time of harvest I will say to the reapers, "First gather together the tares and bind them in bundles to burn them, but gather the wheat into my barn." Matt 13:24-30

When the farm workers found tares (weeds) in the wheat, their response was appropriate. Weeds would grow faster than the wheat, taking nutrients and sun from the crop, then go to seed and thereby multiply the weed problem for years to come. So, no knowledgeable farmer would let the weeds grow with the wheat, spoil the crop, and ruin his field. Why would Jesus advocate that kind of farm policy? He didn't! The story makes His point memorable by clearly contrasting the sound farm policy which the workers knew, with the kingdom of heaven, where God gives probationary time for us to fill our "cups" with good or bad as we wait for the judgement. Jesus had previously taught the same lesson like this:

> "You have heard that it was said, 'You shall love your neighbor and hate your enemy.' But I say to you, love your enemies, bless those who curse you, do good to those who hate you, and pray for those who spitefully use you and persecute you, that you may be sons of your Father in heaven; for **He makes His sun rise on the evil and on the good, and sends rain on the just and on the unjust.** Matt 5:43-46 (emphasis mine).

JESUS INTRODUCES GOD

The Edenic picture of God, The Creator, who formed Adam with His hands, breathed His own life-giving breath into Adam's nostrils, made Eve for his life's companion, and walked with them in the garden, had been lost. Early in His ministry, the gospels record many of the ways that Jesus tried to correct the disciple's image of God. Mathew records one of those ways.

"And when you pray, you shall not be like the hypocrites. For they love to pray standing in the synagogues and on the corners of the streets, that they may be seen by men. Assuredly, I say to you, they have their reward. But you, when you pray, go into your room, and when you have shut your door, pray to your Father who is in the secret place; and your Father who sees in secret will reward you openly. And when you pray, do not use vain repetitions as the heathen do. For they think that they will be heard for their many words. "Therefore do not be like them. For your Father knows the things you have need of before you ask Him. In this manner, therefore, pray:

Our Father in heaven,

Hallowed be Your name.

Your kingdom come.

Your will be done

On earth as it is in heaven.

Give us this day our daily bread.

 And forgive us our debts,

As we forgive our debtors.

And do not lead us into temptation,

But deliver us from the evil one.

For Yours is the kingdom and the power and the glory forever. Amen. Matt 6: 5-14 (emphasis mine)

It is noteworthy that Jesus taught His disciples to address God, as their Father, instead of one of the available names of God at the time; Yahweh, Elohim, or Adonai. Jesus frequent reference to God in the family context, as the head of the human family, was not the model advanced by the clerics of His day. The Father image Jesus advocated evokes a concept of God that makes Him the person we would all like to know.

> Behold what manner of love the Father has bestowed on us, that we should be called **children of God!** Therefore the world does not know us, because it did not know Him. Beloved, now we are children of God; **and it has not yet been revealed what we shall be, but we know that when He is revealed, we shall be like Him,** for we shall see Him as He is. And everyone who has this hope in Him purifies himself, just as He is pure. 1 John 3:1-3 (emphasis mine)

Chapter 10

Evidence Beyond Doubt

Jesus began His public ministry in the town of Capernaum, which was located on the northern shore of the Sea of Galilee. The town was on the North-South trade route between the continents of Asia and Africa; it was also the headquarters for many fishermen. The Sea of Galilee is on the eastern end of the Mediterranean Sea, from which it is separated by a mountain range, whose valleys funnel storms from sea level, down to Galilee, several hundred feet below sea level. The fisherman who made up Jesus's first disciples were well acquainted with the violent storms that come down the mountain valleys from the Mediterranean Sea and turned the Sea of Galilee into a watery grave for many mariners. These men of the sea where on their way back to Capernaum from a missionary journey, when they were confronted with the "out of this world concept" that Jesus had the power and authority to control nature, by verbal command! The narrative goes like this,

> On the same day, when evening had come, He said to them, "Let us cross over to the other side." Now when they had left the multitude, they took Him along in the boat as He was. And other little boats were also with Him. And a great windstorm arose, and the waves beat into the boat, so that it was already filling. But He was in the stern, asleep on a pillow. And they awoke Him and said to Him, "Teacher, do You not care that we are perishing. Then He arose and rebuked the wind

and said to the sea. Peace, be still!" And the wind ceased and there was a great calm. But He said to them, "Why are you so fearful? How is it that you have no faith?" And they feared exceedingly, and said to one another, "Who, can this be, that *even the wind and the sea obey Him!"* Mark 4:35-41 (emphasis mine)

These frightened seamen witnessed Jesus's authority over the material world when they escaped drowning, thanks to the power He demonstrated over the storm. This experience would not be the first or the last time they would witness Jesus's other worldly power. The apostle John would later write that Jesus's exploits could fill many books, but one of the most significant proofs of His divinity would happen just a few days before His death and resurrection. The record goes like this:

> Now a certain man was sick, Lazarus of Bethany, the town of Mary and her sister Martha. It was that Mary who anointed the Lord with fragrant oil and wiped His feet with her hair, whose brother Lazarus was sick. Therefore the sisters sent to Him, saying, "Lord, behold, he whom You love is sick." When Jesus heard that, He said, "This sickness is not unto death, but for the glory of God, that the Son of God may be glorified through it." Now Jesus loved Martha and her sister and Lazarus. So, when He heard that he was sick, He stayed two more days in the place where He was. Then after this He said to the disciples, "Let us go to Judea again." John 11:1-7

Then Jesus explained to His disciples why He had declined the sisters' desperate plea to come and heal Lazarus.

> "Our friend Lazarus sleeps, but I go that I may wake him up."

> Then His disciples said, "Lord, if he sleeps he will get well." However, Jesus spoke of his death, but they thought that He was speaking about taking rest in sleep. Then Jesus said to them plainly, **"Lazarus is dead.** *And I am glad for your sakes that I was not there, that you may believe.* Nevertheless let us go to him." John 11:11-15 (emphasis mine)

Standing at the burial cave, the record says that "Jesus wept",

> Then Jesus, again groaning in Himself, came to the tomb. It was a cave, and a stone lay against it. Jesus said, "Take away the stone." Martha, the sister of him who was dead, said to Him, "Lord, by this time there is a stench, for he has been dead *four days."*

> Jesus said to her, "Did I not say to you that if you would believe you would see the glory of God?" Then they took away the stone from the place where the dead man was lying. And Jesus lifted up His eyes and said, "Father, I thank You that You have heard Me. And I know that You always hear Me, but because of the people who are standing by I said this, that they may believe that You sent Me." Now when He had said these things, He cried with a loud voice, "Lazarus, come forth!" **And he who had died came out bound hand and foot with grave clothes, and his face was wrapped with a cloth. Jesus said to them, "Loose him, and let him go." John 11:38-44** (emphasis mine)

The resurrection of Lazarus is the apex of Jesus's miracles. It documents His authority over the material universe, and His resurrection power of life over death. Like His own upcoming resurrection, that of Lazarus was witnessed by many people, both friend and foe. We can get a little idea about the number of witnesses by listing the category of people present at Lazarus' tomb. There were the sisters of Lazarus, the disciples, the mourners from Jerusalem, the spies from the Jewish authorities (Pharisees and Sadducees), the curious bystanders and the women who followed the disciples and Jesus to make arrangements for food and lodging. The number fifty would likely be a minimum. Of the group who followed Jesus wherever He went, the most significant for this discussion would be the spies. When they reported back to the Jewish leaders, the authorities acted swiftly on what they accepted as reality. Their testimony is very important because it validates the testimony of the disciples. Instead of acting logically based on the evidence, the Jewish leadership sided with Satan.

Then the chief priests and the Pharisees gathered a council and said, "What shall we do? *For this Man works many signs.* If we let Him alone like this, everyone will believe in Him, and the Romans will come and take away both our place and nation." And one of them, Caiaphas, being high priest that year, said to them, "You know nothing at all, nor do you consider that it is expedient for us that *one man should die for the people,* and not that the whole nation should perish." John 11:7-50 (emphasis mine)

Then, from that day on, they plotted to put Him to death. Therefore Jesus no longer walked openly among the Jews, but went from there into the country near the wilderness, to a city called Ephraim, and there remained with His disciples. And the Passover of the Jews was near, and many went from the country up to Jerusalem before the Passover, to purify themselves. Then they sought Jesus, and spoke among themselves as they stood in the temple, "What do you think — that He will not come to the feast?" Now both the chief priests and the Pharisees had given a command, that if anyone knew where He was, he should report it, that they might seize Him. John 11:53-57

His death was now near and Jesus cautioned His followers,

And He strictly warned and commanded them to tell this to no one, saying, "The Son of Man must suffer many things, and be rejected by the elders and chief priests and scribes, and be killed, and be raised the third day." Luke 9:21`-21

The resurrection of Lazarus was/is the validation of Jesus as the Word of God. It happened in broad daylight, witnessed by friends and foes. He had raised the dead previously, but his critics had said He had only resuscitated them. Jewish tradition had established that "corruption" set in dead bodies by the 4th post mortem day. Modern science agrees with that observation, in un-embalmed bodies. The bone cells, which are the last to die, may survive 3days. So, Jesus's stated intent was to permit Lazarus to **completely decay,** so that God's power to verbally CREATE could be demonstrated beyond question.

Chapter 11

A Little Science

It is noteworthy that Jesus was in the grave only 3 days. Lazarus was entombed for at least four days, before being resurrected. The Jewish tradition, based on their experience with unembalmed bodies, was that "corruption" (decay) set in on the 4th day after death. King David had prophesied that the Messiah would not see corruption:

> For You will not leave my soul in Sheol, Nor will You allow Your Holy One to see corruption. Ps 16:10

Jesus had raised dead people before Lazarus, but some detractors, then and now, said that the folks He raised were just asleep or had fainted. Some critics claim that Jesus didn't really die, that He had fainted from the trauma of the cross, and His disciples stole his body from the tomb, revived Him, and claimed that He had risen from the dead. But unlike Jesus, Lazarus was "corrupted" and that is a very significant difference in that no one could claim that his resurrection was a resuscitation. God's ability to **verbally create** was witnessed and documented by friends and enemies alike.

In the 1960's, the invention of cardiac defibrillators, mouth to mouth breathing assistance, and sternal cardiac compression techniques, made it accepted procedure to attempt resuscitation of the recently dead. Indiscriminate resuscitation soon fell into disrepute, however, as it became obvious that some of those resuscitated people, although bodily alive (heartbeat and breathing) were brain dead. It was observed that the cerebral cortex begins to die, if deprived of sugar/and or oxygen, for longer that about 5 minutes. Various other cells in the body die in sequence, starting with the brain at 5 minutes, then the brain stem, then vital organs, ending with the bone cells, which can last up to 72 hours in the un-embalmed human body. (Depending on the temperature of the environment)

The significance of these facts is that when Jesus stood in front of the burial cave where the decayed body of His friend Lazarus lay, the power of His **words** CREATED the man Lazarus in a matter of seconds; memory and all, from nothing alive-*de movo!* This well witnessed account of the verbal **creation** of *the man Lazarus*, even acknowledged by the High Priest, was based on the eyewitness account of the hostile spies, among others, and it confirms the clear assertions of John and Paul in their letters to the churches and to us, in the following quotations.

> In the beginning was the Word, and the Word was with God, and the Word was God. He was in the beginning with God. All things were made through Him, and without Him nothing was made that was made. *In Him was life,* and the life was the light of men. And the light shines in the darkness, and the darkness did not comprehend it. John 1:1-5 (emphasis mine)

> Now faith is the substance of things hoped for, the evidence of things not seen. For by it the elders obtained a good testimony. By faith we understand that the *worlds were framed by the* **Word of God,** so that the things which are seen were not made of things which are visible. Heb 11:1-3 (emphasis mine)

We have the same challenging choice that General Joshua gave the early Israelis in his retirement speech. After sighting their miraculous deliverance from four hundred years of slavery in Egypt, and their recent settlement in the Promised Land, he counsels them to make a decision based on their **experience** and **observation**. Our memory of events, our current experience, and our competent study of the inspired word, are the foundation of FAITH. OBSERVATIONAL SCIENCE relies on the same foundation.

Joshua's counsel;

> "Now therefore, fear the Lord, serve Him in sincerity and in truth, and put away the gods which your fathers served on the other side of the River and in Egypt. Serve the Lord! And if it seems evil to you to serve the Lord, choose for yourselves this day whom you will serve, whether the gods which your fathers served that were on the other side of the River, or the gods of the Amorites, in whose land you dwell. **But as for me and my house, we will serve the Lord."** Josh 24:14-15 (emphasis mine).

CONCLUSIONS

Scientists are entitled to their opinion about anything they wish. They are NOT, however, entitled to their own TRUTH! By its very nature, science and its methodology, is limited to the material creation. The only historically verifiable, coherent information about life, death and eternity is found in the 66 books of the Bible. There we find the reliable testimony of the WORD who came into our material universe to tell us, as He did with Nicodemus, about eternity and the God who made us, loves us in spite of ourselves and redeemed us with the personal sacrifice we can barely understand.

> "For God so loved the world that He gave His only begotten Son, that whoever believes in Him should not perish but have everlasting life. For God did not send His Son into the world to condemn the world, but that the world through Him might be saved." John 3:16

The best description of eternity is found in the last of the 66 books of our Bible, it reads like this,

And he showed me a pure river of water of life, clear as crystal, proceeding from the throne of God and of the Lamb. In the middle of its street, and on either side of the river, was the tree of life, which bore twelve fruits, each tree yielding its fruit every month. The leaves of the tree were for the healing of the nations. And there shall be no more curse, but the throne of God and of the Lamb shall be in it, and His servants shall serve Him. They shall see His face, and His name shall be on their foreheads. There shall be no night there: They need no lamp nor light of the sun, for the Lord God gives them light. And they shall reign forever and ever. Rev 22:1-5

Chapter 12

Question One

William Shakespeare drew our attention to the ultimate question with his memorable words, "To be or not to be, **that** is the question". (Hamlet, Act 3, Scene 1). It is the question that all humans face at least once in our lives. The answer that we choose will have one of two root sources, although there are many variants derived from each root, and some people choose a mixture of the two basic philosophies. The "to be" root is the Judeo- Christian Scriptures, the "not to be" root is the Naturalistic philosophy most often represented by the Darwinian Evolution model.

The Darwinian model embraces the Greek idea that the material universe is all there is. In that system, **Death is necessary** because the less well evolved species must be eliminated in order to provide space and nutrition for the new superior species. This is the "not to be" choice, in which each individual exists **autonomously**, having arrived at life by way of the impersonal laws of nature, expressed without plan or intelligent purpose. Each individual then, having occupied a temporary space in the Evolutionary continuum, passes into oblivion.

The Biblical model presumes, and illustrates, that the material world is a small part of a larger non-material universe that was designed and brought into being by God who lives in Eternity, **outside** of our time-space material universe. From the Biblical perspective each individual was designed and created in perfection with a predetermined form and function. Death, in this scenario, is the result of a catastrophe that was not a part of the original design, although a way to escape the eternal consequences of such an event was provided in the original plan, sometime in eternity. Life was the design and gift of the Designer, Creator, who made us to be part of His family, with personality and character traits like His own, including the ability to choose. We exist **dependent** on our **autonomous Creator.** To live in compliance with the Creator's design is the "to be" choice. In His conversation with Nicodemus, Jesus claimed to represent the "outside of our time-space" universe.

> Jesus answered and said to him, "Are you the teacher of Israel, and do not know these things? Most assuredly, I say to you, We speak what We know and testify what We have seen, and you do not receive Our witness. If I have told you earthly things and you do not believe, how will you believe if tell you heavenly things? No one has ascended to heaven but He who came down from heaven, that is, the Son of Man *who is in heaven.* John 3:10-14 (emphasis mine)

Consider the following narratives from GENESIS:

> Then God said, "Let Us make man in Our image, according to Our likeness; let them have dominion over the fish of the sea, over the birds of the air, and over the cattle, over all the earth and over every creeping thing that creeps on the earth." So God created man in His own image; in the image of God He created him; male and female He created them. Then God blessed them, and God said to them, "Be fruitful and multiply; fill the earth and subdue it; have dominion over the fish of the sea, over the birds of the air, and over every living thing that moves on the earth." Gen 1:26-28

> Then the Lord God took the man and put him in the Garden of Eden to tend and keep it. And the Lord God commanded the man, saying, "Of every tree of the garden you may freely eat; but of the tree of the knowledge of good and evil you hall not eat, for in the day that you eat of it *you shall surely die."* Gen 2:15-17 (emphasis mine)

In the last 150 years, or so, we have seen an increasing number of professed Christians who espouse a mixture of the two root philosophies referred to above. Their co-mingling of opposing philosophies seem to have been driven, in part, by the incredible scientific technology that has transformed our society, and has resulted in, essentially, deifying anything and anybody identified with "science". The consequence has been that many "Christians" espouse some form of the Evolutionary scheme, using the excuse that "we have to believe science". The truth is that the two root philosophies (naturalism and religion) are mutually exclusive. Science is not inerrant and no admixture is coherent.

The studied facts are that an answer to the "death question" is totally outside the competence of science. The Bible is the authoritative and unrivaled source of our information on that topic. Further, it is apparent that the only plausible reason that we know the name of a poor boy from a small, disreputable town in Palestine, who lived and died the death of a criminal, some two thousand years ago, is that He was **resurrected;** the Son of God, the long promised Redeemer (from eternal death) of human kind. Billions of intelligent people have believed the hundreds of credible witnesses, whose testimony we find in the New Testament. Many witnesses, including most of His disciples, accepted a martyrs' death rather than to retract their testimony. An example of their testimony after the resurrection reads like this,

> Now as they said these things, Jesus Himself stood in the midst of them, and said to them, "Peace to you." But they were terrified and frightened, and supposed they had seen a spirit. And He said

to them, "Why are you troubled? And why do doubts arise in your hearts? Behold My hands and My feet, that it is I Myself. Handle Me and see, for a spirit does not have flesh and bones as you see I have." Luke 24:36-39

The apostle Peter would testify to the Jewish leaders,

> This Jesus God has raised up, of which we are all witnesses. Act 2:32-33

The apostle Paul, writing to the Corinthians said,

> Now if Christ is preached that He has been raised from the dead, how do some among you say that there is no resurrection of the dead? But if there is no resurrection of the dead, then Christ is not risen. And if Christ is not risen, then our preaching is empty and your faith is also empty. yes, and we are found false witnesses of God, because we have testified of God that He raised up Christ, whom He did not raise up — if in fact the dead do not rise. For if the dead do not rise, then Christ is not risen. *And if Christ is not risen, your faith is futile;* you are still in your sins! Then also those who have fallen asleep in Christ have **perished.** (They are not in heaven, purgatory or hell!). If in this life only we have hope in Christ, we are of all men the most pitiable. 1 Cor 15:12-19 (emphasis mine)

> For I delivered to you first of all that which I also received: that Christ died for our sins according to the Scriptures, and that He was buried, and that He rose again the third day according to the Scriptures, and that He was seen by Cephas, then by the twelve. After that He was seen by over five hundred brethren at once, of whom the greater part remain to the present, but some have **fallen asleep.** After that He was seen by James, then by all the apostles. Then last of all He was seen by me also, as by one born out of due time. 1 Cor 15:3-8 (emphsis mine)

These are typical of the statements of the New Testament writers, who made the resurrection their key subject. **Resurrection** has been called the single word summary of the New Testament Books. The resurrection idea, however, is also found in the Old Testament. It is the Bible's answer to the death problem. Consider the testimony of Job.

> For I know that my Redeemer lives,
>
> And He shall stand at last on the earth;
>
> And after my skin is destroyed, this I know,
>
> *That in my flesh I shall see God,*
>
> Whom I shall see for myself,
>
> And my eyes shall behold, and not another. How my heart yearns within me! Job 19:25 (emphasis mine)-

Chapter 13

Who Dunnit?

With the invention of the printing press, the telescope and the microscope in the 16th, 17th, and 18th centuries, there came an explosion of information about our world and the universe. Science was born. In prior centuries, many of the diseases that afflicted mankind were attributed to the Devil, or to the sinning he started in the Garden of Eden. We have an example of that idea from Jesus time.

> Now as Jesus passed by, He saw a man who was blind from birth. And His disciples asked Him, saying, "Rabbi, who sinned, this man or his parents, that he was born blind?" John 9:1- 3

As scientific evidence began to identify organisms and other causes of disease, cures soon followed, and over time, as evidence mounted, it seemed that the Devil was not a necessary cause and God was not needed as the necessary cure. As more and more of natural law was discovered, the church was found to be in opposition to some of it. However, at the same time the church was being shown to be corrupt. Martin Luther wrote in his THREE TREATIES about how the church universities of his day had replaced New Testament Pauline teaching with heathen Greek philosophy. The wisdom of Thomas Aquinas was losing its influence on logic and reason.

Between Luther and the present day, the Devil (Satan) has all but disappeared as a real being in many minds. Unable to deny evil in its many venues around the world, some have proposed that God is "testing" His people with temptations to evil! Jesus's brother, James, has some advice for us on that matter:

> Let no one say when he is tempted, "I am tempted by God"; for God cannot be tempted by evil, nor does He Himself tempt anyone. But each one is tempted when he is drawn away by his own desires and enticed. Then, when desire has conceived, it gives birth to sin; and sin, when it is full-grown, brings forth death. James 1:13-15

The Bible adheres to the cause and effect principle as does science. Satan is identified throughout the 66 books as the **cause** of evil and its **effects,** suffering and death. God is our Creator, the life giver, Satan is the inventor of lawlessness, and death. The Bible explains the origin of the good and evil phenomenon like this:

> And war broke out in heaven: Michael and his angels fought with the dragon; and the dragon and his angels fought, but they did not prevail, nor was a place found for them in heaven any longer. So the great dragon was cast out, that serpent of old, called the Devil and Satan, who deceives the whole world; he was cast to the earth, and his angels were cast out with him. Then I heard a loud voice saying in heaven, "Now salvation, and strength, and the kingdom of our God, and the power of His Christ have come, for the accuser of our brethren, who accused them before our God day and night, has been cast down. And they overcame him by the blood of the Lamb and by the word of their testimony, and they did not love their lives to the death. Therefore rejoice, O heavens, and you who dwell in them! Woe to the inhabitants of the earth and the sea! For the devil has come down to you, having great wrath, because he knows that he has a short time." Rev 12:7-12

Sometime in eternity past, Lucifer, The Angel of Light, decided that he, a created being, would be his own god; he tried to over throw his maker by insurrection. He lost and was exiled to the earth where he succeeded in enlisting our first parents in his rebellion.

"How you are fallen from heaven,

O Lucifer, son of the morning!

How you are cut down to the ground,

You who weakened the nations!

For you have said in your heart:

'I will ascend into heaven,

I will exalt my throne above the stars of God;

I will also sit on the mount of the congregation

On the farthest sides of the north;

I will ascend above the heights of the clouds,

I will be like the Most High.'

Yet you shall be brought down to Sheol,

To the lowest depths of the Pit. Isa 14:12-15 NKJV.

Chapter 14

That's How We Got Here

We don't know the condition of the earth when Satan was exiled here. We know that Adam had named the animals, met and married Eve and took evening walks with the Creator. That was before they met Satan in his animal disguise. It is speculation as to what the Creator told our parents about the war in Heaven and His enemy, but there is no doubt in my mind that they were told enough to be able to make the right choice. They didn't.

> The Lord God planted a garden eastward in Eden, and there He put the man whom He had formed. And out of the ground the Lord God made every tree grow that is pleasant to the sight and good for food. The tree of life was also in the midst of the garden, and the tree of the knowledge of good and evil. Gen 2:8-9

> Then the Lord God took the man and put him in the Garden of Eden to tend and keep it. And the Lord God commanded the man, saying, "Of every tree of the garden you may freely eat; but of the tree of the knowledge of good and evil you shall not eat, for in the day that you eat of it you shall surely die." Gen 2:15-17

Sooner or later, we humans are led to ponder life's ultimate question, why do we die? The usual occasion for the query is the death of something or someone loved. Funerals are such an occasion. If a person subscribes to the Evolutionary scheme of origins, the obvious answer to the question is that the plants and animals not yet perfected must die in order to make space and nutrients for their more perfect successors.

Alternatively, some "believers" in the Biblical Creation narrative have suggested that The Creator, in essence, kills His disobedient/rebellious creatures. They believe that God required the death of Jesus to satisfy His need or desire for justice and to save humanity from its just reward.

If either of these explanations fails to satisfy, the diligent student of Scripture will find the historic evidence that reveals the real reason we die. The Creator, in His text book (Scripture), has given us the answer to this important question, and we can find that the answer is neither of those suggested above. Another idea, borrowed from pagan religions, is that we don't really die! They say that we just transition to Heaven or Hell. Seems to me that was Satan's line to Eve, "You shall not surely die".

In his letter to the Roman Christians, Paul identifies "sin" (Satan's invention) as the cause of death, and contrasts it with "life" which the Creator God imparts to His Creation. So, in modern scientific, cause and effect lingo, the cause of death is "sin", and the cause of life is the Creator God. God and imparted life are the alternatives to Satan's sin and death.

The NKJV of Scripture has 466 references to sin, and 199 references to sins. The importance of the subject in Scripture is obvious. Although both words are used to describe the actions that eventually result in death, the singular word sin is also used to describe a state of being, a *mind-set(heart)*, a state of rebellion which is the cause of the "death actions." So in order to understand the acts of sin, it is necessary to understand the *"state of mind"* that results in the sins. Jesus explained it like this,

> "But those things which proceed out of the mouth come from the heart, and they defile a man. For out of the heart (mind) proceed

evil thoughts, murders, adulteries, fornications, thefts, false witness, and blasphemies. These are the things which defile a man, but to eat with unwashed hands does not defile a man." Matt 15:18-20 NKJV.

Therefore, in order to understand the origin of the sin problem and the death that has resulted, we need to understand the original *mindset* that is the *cause* of the sin *effect*. For that answer we need to go back to the very earliest recorded sin, so we can discover the nature of the problem. Isaiah gives us the history we need.

"How you are fallen from heaven,

O Lucifer, son of the morning!

How you are cut down to the ground,

You who weakened the nations!

For you have said in your heart:

'I will ascend into heaven,

I will exalt my throne above the stars of God;

I will also sit on the mount of the congregation

On the farthest sides of the north;

I will ascend above the heights of the clouds,

I will be like the Most High.' Isa 14:12-14 NKJV. (emphasis mine)

The Apostle John supplies us with additional details of the sin story, titled,

Satan Thrown out of Heaven.

"And war broke out in heaven: Michael and his angels fought with the dragon; and the dragon and his angels fought, but they did not prevail, nor was a place found for them in heaven any longer. So the great dragon was cast out, that serpent of old, called the Devil and Satan, who deceives the whole world; *he was cast to the earth*, and his angels were cast out with him." Rev 12:7-9 NKJV. (emphasis mine)

There are some further details of Lucifer's (Satan's) rebellion,

"I established you; you were on the holy mountain of God; You walked back and forth in the midst of fiery stones. You were perfect in your ways from the day you were created, Till iniquity was found in you. By the abundance of your trading You became filled with violence within, And **you sinned;** Therefore I cast you as a profane thing Out of the mountain of God; And I destroyed you, O covering cherub, From the midst of the fiery stones. Your heart was lifted up because of your beauty; *You corrupted your wisdom* for the sake of your splendor; *I cast you to the ground,...*" Ezek 28:14-18 NKJV. (emphasis mine)

Our first meeting of the *exiledSatan* is in his conversation with Eve in the Garden of Eden, where he appears in disguise. "Now the serpent was more cunning than any beast of the field which the Lord God had made. And he said to the woman,

"Has God indeed said, 'You shall not eat of every tree of the garden'?" And the woman said to the serpent, 'We may eat the fruit of the trees of the garden; but of the fruit of the tree which is in the midst of the garden, God has said, 'You shall not eat it, nor shall you touch it, lest you die.' Then the serpent said to the woman, *'You will not surely die.* For God knows that in the day you eat of it your eyes will be opened, and ***you will be like God,*** knowing good and evil.'" Gen 3:1-5 (emphasis mine)

So, it was that the sales pitch Satan brought to Eve was a re-statement of his declaration of **autonomy** in Paradise; **"I will be like the Most High."** Lucifer and Eve were both created beings who by virtue of their creation were in some respects "[L]ike the Most High." ("Then God said, 'Let Us make man in Our image, according to Our likeness;'" Gen1:26). But the Satanic assertion that he and humans could be autonomous (knowing good and evil) and thus equal to and independent of their Creator, was absurd in the extreme, and so unreasonable as to qualify as "the *mystery of iniquity.*" Lucifer, by his attempt to be equal to and independent from his own creator, succeeded in making himself into Satan, the inventor of the sin mind set, the father of lies, the great deceiver. His great lie to himself, Eve and others, was that we creatures can be **autonomous** and free of the obligations inherent in our *design*. (If Satan was a man, he might drive a two tone, straight eight, 4-door Pontiac right into the ground.) The purpose of God's forbidden tree in Eden was to give our parents the power to choose, and to illustrate that created beings are **limited** by virtue of their *design and creation*. The life of created beings is imparted and sustained by their Creator. The fourth commandment of the Decalogue makes clear that we are **dependent** on God for both our *creation and our sustenance*.

Jesus's answer to one of Satan's wilderness temptations is a fitting summary of the dependence/autonomy issue,

> "Then Jesus was led up by the Spirit into the wilderness to be tempted by the devil. And when He had fasted forty days and forty nights, afterward He was hungry. Now when the tempter came to Him, he said, 'If You are the Son of God, command that these stones become bread.' But He answered and said, *'It is written, 'Man shall not live by bread alone, but by every word that proceeds from the mouth of God.'"* Matt 4:1-4 (emphasis mine)

The one word summary of our existence is that we creatures are DEPENDANT. Thank God for all His gifts. John tells us,

> "And he showed me a pure river of water of life, clear as crystal, proceeding from the throne of God and of the Lamb. In the middle

of its street, and on either side of the river, was **the tree of life**, which bore twelve fruits, each tree yielding its fruit every month. The leaves of the tree were for the healing of the nations." Rev 22:1-3 (emphasis mine)

It seems that our eternal existence, even in the presence of God, may be linked to the Tree of Life, as it was in Eden?

But, there is more. The Creator not only gave us His "likeness", He gave our parents earthly **"dominion"**. Adam and Eve were to be the rulers of the earth and the progenitors of its population.

"Then God said, 'Let Us make man in Our image, according to Our likeness; let them have **dominion** over the fish of the sea, over the birds of the air, and over the cattle, over all the earth and over every creeping thing that creeps on the earth.' So God created man in His own image; in the image of God He created him; male and female He created them. Then God blessed them, and God said to them, 'Be fruitful and multiply; fill the earth and subdue it; have **dominion** over the fish of the sea, over the birds of the air, and over every living thing that moves on the earth.'" Gen 1:26-28 (emphasis mine)

God gave our first parents life, **dominion** of the earth and the ability to share in His creative nature by populating the planet. By siding with Satan's autonomy scheme, however, *they cut themselves off* from the Life Giver and aligned themselves with God's enemy. Satan then claimed them as his subjects, in life and in death, and assumed their God given **dominion** as his own. Consider the evidence,

"Now there was a day when the Sons of God came to present themselves before the Lord, and Satan also came among them. And the Lord said to Satan, 'From where do you come?' So Satan answered the Lord and said, 'From going to and fro on the earth, and from walking back and forth on it.'" Job 1:6-7

We find another of Satan's claims to dominion of the earth and its inhabitants in the book of Jude.

"Yet Michael the archangel, in contending with the devil, when **he disputed about the body of Moses,** dared not bring against him a reviling accusation, but said, 'The Lord rebuke you!'" Jude 9-10 (emphasis mine)

And we have the most absurd claim of all, in Satan's assertion of **his earthly dominion** when he made his proposition to Jesus, the Creator, in person,

"Again, the devil took Him up on an exceedingly high mountain, and showed Him *all the kingdoms of the world* and their glory. And he said to Him, 'All these things *I will give* You if You will fall down and **worship me**.'" Matt 4:8-9 (emphasis mine)

In an Old Testament prophesy we find Christ's remedy to Satan's assertion of **dominion,**

"I will **ransom** them from the power of the grave;

I will **redeem** them from death.

O Death, I will be your plagues!

O Grave, I will be your destruction!" Hos 13-14 (emphasis mine)

The fulfillment of that prophesy and the definitive answer to the **dominion** issue, we find in Christ's own words,

"Just as the Son of Man did not come to be served, but to serve, and to give His life a **ransom** for many." Matt 20-23 (emphasis mine)

The poetry of a favorite old hymn addresses this profound mystery:

Jesus keep me near the cross

There a precious fountain'

Comes to all a healing stream

Flows from Calvary's mountain... till my **ransomed** soul shall find, rest beyond the river. (emphasis mine)

Our world is full of cemeteries which keep getting larger, the result of the funerals of friends, enemies and billions of people who we will never know. The incontrovertible evidence is that God told Eve the truth. Satan continues his lying, and people keep dying! Our faith in God and His word is based on the "substance and evidence" that the historical facts engender. See Hebrews 11:1

Here is the Christian's answer to the death problem,

> "For God so loved the world that He gave His only begotten Son, that whoever believes in Him should not perish but have **everlasting life.** For God did not send His Son into the world to condemn the world, but that the world through Him might be saved." John 3:16 (emphasis mine)

Then there is this:

> "There is therefore **now** no condemnation to those who are in Christ Jesus, who do not walk according to the flesh, but according to the Spirit. For the law of the Spirit of life in Christ Jesus has made me *free from the law of sin and death."* Rom 8:1-3 (emphasis mine)

And finally, we have Paul's summary of life's reality to the Greeks, and to us,

> "God, who made the world and everything in it, since He is Lord of heaven and earth, does not dwell in temples made with hands. Nor is He worshiped with men's hands, as though He needed anything, since **He gives to all life, breath, and all things.** And He has made from one blood every nation of men to dwell on all the face of the earth, and has determined their pre-appointed times and the boundaries of their dwellings, so that they should seek the Lord, in the hope that they might grope for Him and find Him, though He is not far from each one of us; *for in Him (alternative translation: "because of Him")we live and move and have our being, as also some of your own poets have said, 'For we are also His offspring.'"* Acts 17:24-29 (emphasis mine).

God the Creator is the Devine Parent we inherently seek. That is why Jesus begins His model prayer with the words, "Our Father...".

Chapter 15

Two Old Trees

The names of the two oldest trees on record are found in Genesis chapters 1-3.

> The Lord God planted a garden eastward in Eden, and there He put the man whom He had formed. And out of the ground the Lord God made every tree grow that is pleasant to the sight and good for food. Gen 2:8-9

> Then the Lord God took the man and put him in the Garden of Eden to tend and keep it. And the Lord God commanded the man, saying, "Of every tree of the garden you man freely eat; but of the tree of the knowledge of good and evil you shall not eat, for in the day that you eat of it you shall surely die" Gen 2:15-17

> So when the woman saw that the tree was good for food, that it was pleasant to the eyes, and a tree desirable to make one wise, she took of its fruit and ate. She also gave to her husband with her, and he ate. Then the eyes of both of them were opened, and they knew that they were naked; and they sewed fig leaves together and make themselves coverings. Gen 3:6-7

> Then, the Lord God said, "behold, the man has become like one of Us, to know good and evil. And now, lest he put out his hand and take also of the tree of life, and eat, and live forever"-therefore the Lord God sent him out of the Garden of Eden to till the ground

> from which he was taken. So He drove out the man; and He placed cherubim at the east of the Garden of Eden, and a flaming sword which turned every way, to guard the way to the tree of life. Gen 3:22-24

An analysis of these very significant old trees gives us the foundation we need to achieve to understand the history of mankind. We are able to construct a realistic world view built on the foundation laid in Genesis and expanded in the 65 books that follow it. The result of such an exercise can be a rewarding understanding of the relationship that the Creator desires with His creation through an understanding of the **Law of Design,** and its companion, the **Law of Being.** That understanding is the key to unlock the Mystery of Iniquity and the remedial Gospel of Christ. Let's start our analysis with the conversation between Eve and Satan at the forbidden tree.

> Now the serpent was more cunning than any beast of the field which the Lord God had made. And he said to the woman, "Has God indeed said, 'you shall not eat of every tree of the garden'?" And the woman said to the serpent, "We may eat the fruit of the trees of the garden; but of the fruit of the tree which is in the midst of the garden, God has said, 'you shall not eat it, nor shall you touch it, lest you die.'" Then the serpent said to the woman, "You will not surely die. God knows that in the day you eat of it your eyes will be opened, and you will be like God, knowing good and evil.'" Gen 3:1-3

Satan's allegation, is on its face ridiculous. It is clearly not possible for created beings to be equal to their Creator. The Creator's design is His gift to the creature, and the capacity and capability He imparts constitutes their Law of Design and Being, the precepts of which defined the conditions for existence and the very nature of the created being. The eternal, self-existent Creator alone is AUTONOMOUS, and all His creation is by nature DEPENDANT.

The knowledge of the Creator must always exceed that of His creation. Satan's allegation to the contrary, marks the beginning of an evolution from his created identity as Lucifer, the light bearer to Satan, the father of lies. Scripture calls this rebellion "the mystery of iniquity" because it was completely unreasonable.

> For thou hast said in thine heart, I will ascend into heaven, I will exalt my throne above the stars of God: I will sit also upon the mount of the congregation, in the sides of the north: I will ascend above the heights of the clouds; **I will be like the most High.** Isa 14:13-14 (emphasis mine).

The first tree (knowledge of good and evil) confronts us with the truth that we are created, dependent creatures, given life designed by the Creator God whose home is eternity past, present and future. We owe Him for our life and our sustenance. Human kind did not exist until God purposed to share His life with us. The tree gave our parents the choice to respect the Creator, or not.

The Tree of Life taught our first parents their continuing dependence on God for the life He gives. The expulsion from Eden was result of the willful separation of Adam and Eve from their source of life and sustenance.

> So He drove out the man; and He placed cherubim at the east of the Garden of Eden, and a flaming sword which turned every way, to guard the way to the tree of life. Gen 3:24

New Testament history tell us about Jesus' institution of a memorial to the principles illustrated by the old trees.

> And He took bread, gave thanks and broke it, and gave it to them , saying, "This is My body which is given for you; do this in remembrance of Me. Likewise He also took the cup after supper, saying, " This cup is the new covenant in My blood, which is shed for you." Luke 22:19-21.

Life (blood) and its sustenance (bread) are the symbols that Jesus used in the communion ordinance to remind us of His gift of life and our dependence on Him for sustenance.

> Whosoever eats my flesh and drinks My blood has eternal life, and I will raise him up at the last day. For My flesh is food indeed, and My blood is drink indeed. He who eats My flesh and drinks My blood abides in Me, and I in him. John 6:54-57

Paul's version of the "two trees" principle provides another helpful prospective. In his address to the Athenians he said,

> For in Him we live and move and have our being, as also some of your own poets have said, "for we are also His offspring." Acts 17:28-29

Satan's apparent victory over Christ on the cross, demonstrated to the universe that he, the inventor of death, had lied to our parents. Satan lied and God, the Creator, died. On Easter morning God confirmed Himself the truth-teller, the giver and sustainer of life. Our faith in Christ now supplants our need for the first tree. The second old tree was taken to eternity and we find it located in a very prominent place.

> And He showed me a pure river of water of life, clear as crystal, proceeding from the throne of God and of the Lamb. In the middle of its street, and on either side of the river, was the tree of life, which bore twelve fruits, each tree yielding its fruit every month. The leaves of the tree were for the healing of the nations. Rev 22:1-2

> Blessed are they that do His commandments that they may have right to the tree of life, and may enter in through the gates into the city". Rev 22:14

Chapter 16

Jesus And Genesis

It is not uncommon for Christians to say that the Old Testament is not as relevant to their faith in Jesus as the New Testament. While it is true that the Old Testament is largely historical, its importance is too often underestimated. Even the dreaded genealogies may seem irrelevant, until we read Mathew's opening chapter. Unfortunately, many Christians have felt the need to accommodate the Darwinian pseudo-science of earths beginning, at the expense of Genesis credibility. So let's see what Jesus thought about Genesis and our Old Testament.

> You search the Scriptures, for in them you think you have eternal life; and these are they which testify of Me. John 5:39-40

> And Jesus answered and said to them, "Because of the hardness of your heart he wrote you this precept. But from **the beginning of the creation, God 'made them male and female.'** 'For this reason a man shall leave his father and mother and be joined to his wife, and the two shall become one flesh'; so then they are no longer two, but one flesh. Therefore what God has joined together, let not man separate." Marl 10:5-9 (emphasis mine)

> "But of that day and hour no one knows, not even the angels of heaven, but My Father only. But as the days of Noah were, so also will the coming of the Son of Man be. For as in the days before

the flood, they were eating and drinking, marrying and giving in marriage, until the day that **Noah entered the ark,** and did not know until the flood came and took them all away, so also will the coming of the Son of Man be. Matt 24:36-40 (emphasis mine)

So they said to Him, "The things concerning Jesus of Nazareth, who was a Prophet mighty in deed and word before God and all the people, and how the chief priests and our rulers delivered Him to be condemned to death, and crucified Him. But we were hoping that it was He who was going to redeem Israel. Indeed, besides all this, today is the third day since these things happened. Yes, and certain women of our company, who arrived at the tomb early, astonished us. When they did not find His body, they came saying that they had also seen a vision of angels who said He was alive. And certain of those who were with us went to the tomb and found it just as the women had said; but Him they did not see." Then He said to them, "O foolish ones, and slow of heart to believe in all that the prophets have spoken! Ought not the Christ to have suffered these things and to enter into His glory?" **And beginning at Moses and all the Prophets, He expounded to them in all the Scriptures the things concerningHimself. Luke 24:19-27** (emphasis mine)

Chapter 17

Bibical Faith or Presumption

In a sizeable protestant church in northwest Pennsylvania, the associate pastor, a recent seminary graduate, presented the morning sermon titled "faith". He constructed a theoretical scenario to illustrate his idea. Faith, he said, is like climbing to the top floor of a tall building on a dark, moonless night and jumping off into the pitch blackness, not knowing whether you will live or die! Faith, he said, is believing the unbelievable, trusting something or someone against all reason.

There were many blank stares as the congregants looked around to see if others had heard what they heard. The church, located in a college town, had more than its share of folks who recognized his example as the classic definition of **presumption, the polar opposite of faith.** Some wondered if this associate pastor thought the congregation had checked their brains at the door. Did he learn that in seminary?

In chapter eleven of the book or Hebrews, perhaps authored by the Apostle Paul, we find the **definitive definition of faith,** followed by examples.

> Now faith is the **substance** of things hoped for, the **evidence** of things not seen. For by it the elders obtained a good testimony. Heb 11:1-2

By faith we understand that the worlds were framed by the word of God, so that the things which are seen were not made of things which are visible. Heb 11:3

By faith Enoch was taken away so that he did not see death, "and was not found, because God had taken him"; for before he was taken he had this testimony, that he pleased God. But without faith it is impossible to please Him, for he who comes to God must believe that He is, and that He is a rewarder of those who diligently seek Him. Heb11: 5-6

Joshua, Moses' successor, put his faith in God, who had miraculously lead Israel from slavery to freedom; using that historical fact as the reason for his choice.

Now therefore, fear the Lord, serve Him in sincerity and in truth, and put away the gods which your fathers served on the other side of the River and in Egypt. Serve the Lord! And if it seems evil to you to serve the Lord, choose for yourselves this day whom you will serve, whether the gods which your fathers served that were on the other side of the River, or the gods of the Amorites, in whose land you dwell. But as for me and my house, we will serve the Lord." Josh 24: 14-15

In the Old Testament, King David writes about the evidence for his faith.

The heavens declare the glory of God;

And the firmament shows His handiwork.

Day unto day utters speech,

And night unto night reveals knowledge.

There is no speech nor language

Where their voice is not heard.

Their line has gone out through all the earth,

And their words to the end of the world.

In them He has set a tabernacle for the sun,

Which is like a bridegroom coming out of his chamber,

And rejoices like a strong man to run its race.

Its rising is from one end of heaven,

And its circuit to the other end;

And there is nothing hidden from its heat. Ps 19:1-6

In the New Testament, Paul puts it this way:

For the wrath of God is revealed from heaven against all ungodliness and unrighteousness of men, who suppress the truth in unrighteousness, because what may be known of God is manifest in them, for God has shown it to them. For since the creation of the world His invisible attributes are clearly seen, being understood by the things that are made, even His eternal power and Godhead, so that they are without excuse, because, although they knew God, they did not glorify Him as God, nor were thankful, but became futile in their thoughts, and their foolish hearts were darkened. Professing to be wise, they became fools, and changed the glory of the incorruptible God into an image made like corruptible man — and birds and four-footed animals and creeping things. Rom 1:18-23

Chapter 18

Miracles

In the big city hospital where I spent my senior year in medical school, we had lectures from medical clinicians, clergy and an occasional attorney. They shared some valuable lessons they had learned, sometimes the hard way, in order to help us avoid similar situations.

The hospital was located down town, next to a big parking garage. The physicians using the garage would come to park, on the one-way street that led to the entry. Some neer-do-wells had devised an income stream which they derived from auto insurance companies. They would stand close to the curb where the doctors turned to the parking entry, and as the car turned the corner they would fall out into its path, claiming injury. Some of the insurance companies would settle for a nominal amount of money, to avoid the greater expense of a law suit.

In one particularly egregious case that had gone to trial, the jury had awarded a very generous amount of money to the pseudo injured plaintiff. As the frustrated defense attorney left the courtroom, he told the plaintiff that a photographer would be following him everywhere he went, and if he ever left his wheel chair, it would be on film, for the court to review.

The smiling plaintiff responded by taking an envelope from his coat pocket, waving it in the attorneys face, and saying that's a great idea. This envelope has my ticket to the Lourdes Shrine, in the French Pyrenees, where you will be able to document the greatest miracle you have ever seen. My wheel chair will stay at the Shrine, and I will be flying home first class. You are welcome to go with me. Our class smiled at the ingenuity of the crook, but we understood the attorneys warning to us and shared his frustration. It seems that anytime there is something good, someone comes up with a counterfeit.

Some years later while traveling in Europe, I had occasion to see several town markets where display cases were full of "Rolex" watches, at bargain basement prices. They were all fake.

Many years later, I joined a Radiology practice in a large community hospital that had just installed a CT scanner. One morning there was some excitement in the department when a small crowd of doctors and nurses from the adjacent Emergency Room, brought an auto accident victim for a CT scan of the head. A little while later, the CT technician brought me the films with the small crowd of doctors and nurses, to see the results.

The patient was the Director of the local nursing school, well known to many of the hospital staff. I had not known her, being a recent addition to the staff. The patient had undergone head trauma when her car left the interstate 79 at highway speed and encountered a large tree. She was alone. The car was totaled. She was unresponsive. The CT scan showed no bleeding, so no surgery was indicated. The patient was admitted to the hospital and after being unresponsive for weeks, she was transferred to a long term care facility, where she had a feeding tube and general maintenance care.

The accident happened sometime in November, when deer hunting season was in full swing, and the deer were on the move. People thought that the Director had lost control because she was trying to avoid hitting a deer. That was not an uncommon cause of accidents every fall in deer country.

The Director remained unresponsive, in a deep coma, for months. Then in about March, she was brought to the department for a follow-up CT head scan. The result, compared to the original scan made some of her many friends cry. The cortical atrophy was appalling. There was talk now about pulling the feeding tube so that she could complete the dying process that was apparently well underway.

It was then that I learned, from department personnel that beginning with her hospitalization, the nurses in our hospital had started morning and evening prayer sessions, and that there were similar prayer groups at her church, the nursing school and the other hospital in town. The prayer groups continued after the second scan. The family elected to continue her nutrition.

Some few weeks later, about mid-morning, there was another commotion in the department, and one of the technicians rushed to tell me the following story: A nurse care-giver was giving the Director her morning bath, when she opened her eyes and asked; "where is my husband?" Pretty close to a resurrection!

That would have been about March. I attended the nurses' graduation banquet in May of that year, and to my great surprise found my name card directly across the table from the Director. She was lovely, conversant, happy and healthy looking. She had resumed her duties as the Director of nurses and she continued for some years afterward. The incident reminded me of something Jesus's brother wrote.

> Is anyone among you suffering? Let him pray. Is anyone cheerful? Let him sing psalms. Is anyone among you sick? Let him call for the elders of the church, and let them pray over him, anointing him with oil in the name of the Lord. **And the prayer of faith will save the sick, and the Lord will raise him up.** And if he has committed sins, he will be forgiven. Confess your trespasses to one another, and pray for one another, that you may be healed. **The effective, fervent prayer of a righteous man avails much.** Elijah

was a man with a nature like ours, and he prayed earnestly that it would not rain; and it did not rain on the land for three years and six months. And he prayed again, and the heaven gave rain, and the earth produced its fruit.

James 5:13-18 (emphasis mine)

Chapter 19

How Big?

The professor of Old Testament opened his class with the question, "How big is your God?" He was met with frowns and questioning looks. He apparently thought it was a clever way of checking to see if the class had done his assigned reading in the book of Job.

Job is thought by many Bible scholars to be the oldest of the books that make up our Bible; even older than the Genesis record, the later which was likely written by Moses during the 40 year trek through the desert, from Egypt to Canaan.

Moses was self-exiled in Midian after killing an Egyptian slave master for abusing a fellow Abrahamic descendant. While there he encountered the Lord in the burning bush, married his wife, and likely learned about and wrote about Job. From there, the Lord sent him back to Egypt to lead the enslaved descendants of Abraham to the Promised Land.

The book contains conversations between Job and his friends, Job and God, and God's answers to Job. God's questions to Job not only give us a great insight into His creative ability, but also help us answer the professor's question.

Then the Lord answered Job out of the whirlwind, and said:

"Who is this who darkens counsel By words without knowledge? Now prepare yourself like a man; I will question you, and you shall answer Me.

"Where were you when I laid the foundations of the earth? Tell Me, if you have understanding. Who determined its measurements? Surely you know! Or who stretched the line upon it? To what were its foundations fastened? Or who laid its cornerstone, When the morning stars sang together, And all the sons of God shouted for joy?

"Or who shut in the sea with doors, When it burst forth and issued from the womb; When I made the clouds its garment, And thick darkness its swaddling band; When I fixed My limit for it, And set bars and doors; When I said, 'This far you may come, but no farther, And here your proud waves must stop!'

"Have you commanded the morning since your days began, And caused the dawn to know its place, That it might take hold of the ends of the earth, And the wicked be shaken out of it? It takes on form like clay under a seal, And stands out like a garment. From the wicked their light is withheld, And the upraised arm is broken.

"Have you entered the springs of the sea? Or have you walked in search of the depths? Have the gates of death been revealed to you? Or have you seen the doors of the shadow of death?

Have you comprehended the breadth of the earth? Tell Me, if you know all this.

"Where is the way to the dwelling of light? And darkness, where is its place, That you may take it to its territory, That you may know the paths to its home? Do you know it, because you were born then, Or because the number of your days is great?

"Have you entered the treasury of snow, Or have you seen the treasury of hail, Which I have reserved for the time of trouble , For the day of battle and war? By what way is light diffused, Or the east wind scattered over the earth?

"Who has divided a channel for the overflowing water, Or a path for the thunderbolt, To cause it to rain on a land where there is no one, A wilderness in which there is no man; To satisfy the desolate waste, And cause to spring forth the growth of tender grass? Has the rain a father? Or who has begotten the drops of dew? From whose womb comes the ice? And the frost of heaven, who gives it birth? The waters harden like stone, And the surface of the deep is frozen.

"Can you bind the cluster of the Pleiades, Or loose the belt of Orion? Can you bring out Mazzaroth in its season? Or can you guide the Great Bear with its cubs? Do you know the ordinances of the heavens? Can you set their dominion over the earth?

"Can you lift up your voice to the clouds, That an abundance of water may cover you? Can you send out lightnings, that they may go, And say to you, 'Here we are!'? Who has put wisdom in the mind? Or who has given understanding to the heart? Who can number the clouds by wisdom? Or who can pour out the bottles of heaven, When the dust hardens in clumps, And the clods cling together?

"Can you hunt the prey for the lion, Or satisfy the appetite of the young lions, When they crouch in their dens, Or lurk in their lairs to lie in wait? Who provides food for the raven, When its young ones cry to God, And wander about for lack of food?

God Continues to Challenge Job

"Do you know the time when the wild mountain goats bear young? Or can you mark when the deer gives birth? Can you number the months that they fulfill? Or do you know the time when they bear young? They bow down, They bring forth their young, They deliver their offspring. Their young ones are healthy, They grow strong with grain; They depart and do not return to them.

"Who set the wild donkey free? Who loosed the bonds of the onager, Whose home I have made the wilderness, And the barren land his dwelling? He scorns the tumult of the city; He does not heed the shouts of the driver. The range of the mountains is his pasture, And he searches after every green thing.

"Will the wild ox be willing to serve you? Will he bed by your manger? Can you bind the wild ox in the furrow with ropes? Or will he plow the valleys behind you? Will you trust him because his strength is great? Or will you leave your labor to him? Will you trust him to bring home your grain, And gather it to your threshing floor?

"The wings of the ostrich wave proudly ,But are her wings and pinions like the kindly stork's? For she leaves her eggs on the ground, And warms them in the dust; She forgets that a foot may crush them, Or that a wild beast may break them. She treats her young harshly, as though they were not hers; Her labor is in vain, without concern, Because God deprived her of wisdom, And did not endow her with understanding. When she lifts herself on high, She scorns the horse and its rider.

"Have you given the horse strength? Have you clothed his neck with thunder? Can you frighten him like a locust? His majestic snorting strikes terror. He paws in the valley, and rejoices in his strength; He gallops into the clash of arms. He mocks at fear, and is not frightened; Nor does he turn back from the sword. The quiver rattles against him, The glittering spear and javelin. He devours the distance with fierceness and rage; Nor does he come to a halt because the trumpet has sounded. At the blast of the trumpet he says, 'Aha! 'He smells the battle from afar, The thunder of captains and shouting.

"Does the hawk fly by your wisdom, And spread its wings toward the south? Does the eagle mount up at your command, And make its nest on high? On the rock it dwells and resides, On the crag of the rock and the stronghold. From there it spies out the prey; Its eyes observe from afar. Its young ones suck up blood; And where the slain are, there it is." Job 38 and 39

We find additional testimonials to the majesty of God in the New Testament. The apostle Paul describes an experience in what he calls the "third heaven" which was beyond his ability to describe.

It is doubtless not profitable for me to boast. I will come to visions and revelations of the Lord: I know a man in Christ who fourteen years ago — whether in the body I do not know, or whether out of the body I do not know, God knows — such a one was caught up to the third heaven. And I know such a man — whether in the body or out of the body I do not know, God knows — how he was caught up into Paradise and heard inexpressible words, which it is not lawful for a man to utter. 2 Cor 12:1-5

In the same letter to the Corinthians, he adds his interpretation of Isa. 64:4,

"Eye has not seen, nor ear heard,

Nor have entered into the heart of man

The things which God has prepared for those who love Him."

1 Cor 2:9

When the apostle John was shown a vision of the New Jerusalem, saw things he had never seen, and had only earthly words to describe his vision. He had to say that the streets were made of transparent gold, and the gates were carved from a pearl. (Some oyster) How hard is it describe what you have never before seen? It's a little like the problem we humans have trying to get the concept of eternity in our head. There aren't enough circuits in our heads to contemplate eternity past, present and future. Being creatures of time, we find the idea of timelessness a challenging concept.

Chapter 20

Mystery And Humility

In the movie SOUND OF MUSIC, Julie Andrews, having finally admitted to herself that she is in love with the Captain and his children, sings a love song which has the words, "Nothing comes from nothing, nothing ever could. So somewhere in my wicked, miserable past, I must have done something good." Having had the opportunity of traveling in the Swiss Alps, the Rocky Mountains, the great American desert and over the boundless oceans, camping under starry skies and staring into the depths of space, I have often faced the question, who did this? The immediate, obvious answer is, "Not me!" That's evidentiary humility! I agree with the writer of Julie's song that somebody did it. It had a cause. That's science.

The deniers of the Creator God of the Bible have come up with an explanation of how our world and universe came to be. They call their theory "the big bang". The evidence they cite is based on their observation that the universe we can see, seems to be expanding. So, looking backwards in time, the universe would perhaps have been infinitely small at its beginning. The theory purports to explain how we came to be, but it fails the most fundamental scientific requirement. What or (forbid the thought, who) was the cause? "O", they say, the cause was likely a magnetic flux. But, as Julie said, "Nothing comes from nothing, nothing ever could." So, where did the flux come from?

The answer magnetic flux idea reminds me of a statement by the great scientist, professed atheist, co-discoverer of the DNA structure, Francis Crick. He was so impressed with the complex structure of DNA, that he broke with his fellow evolutionists, and said that DNA could not possibly be the result of chance. Life, he said, must have been brought to earth by aliens! From his elegant science, he jumped to an outrageous conclusion, in one giant bound. (Sorry Dr. Smith)

The alien(The only documented alien is Jesus!) argument introduces the concept of an infinite regression, whereby the advocate can postulate, as in this case, that life was carried from one alien to another, backwards into infinity, thus avoiding the question as to where and when life began. Science and classic scientism on displayed here.

Enter the humble, truth seeking scientist with a big enough intellect to say, "I don't know." Why not consider the cause (Creator) and effect (observed universe) answer? Maybe somebody else has an evidentiary answer, Dr. Crick. How many aliens have you interviewed? The genuine searching intellect needs to have a place in the brain to store unexplained findings, while the search for the cause continues. It's where the humble scientist keeps his/her mysteries. Most of us know a little about a lot of things, and nobody knows everything about anything. That doesn't stop us from working within the bounds of our partial knowledge, but should keep us humble.

My curiosity about magnets and magnetism once led me to consult the ENCYCLOPEDIA BRITANNICA, in order to try to understand what it was. I waded through the multipage formulas and equations, not understanding much of it. Some very smart people had expressed their findings in terms of math and physics equations telling **what it does, but not what it is.** Ditto light. To be filed under mysteries for now.

Religion also has its mysteries. How is it that Lucifer, light bearer, created perfect, could become the "dark" bearer, purveyor of death and father of lies?

Paul, in his letter to the Roman church, introduces us to the other Biblical mystery.

> For when we were still without strength, in due time Christ died for the ungodly. For scarcely for a righteous man will one die; yet perhaps for a good man someone would even dare to die. **But God demonstrates His own love toward us, in that while we were till sinners, Christ died for us.** Much more then, having now been justified by His blood, we shall be saved from wrath through Him. For if when we were enemies we were reconciled to God through the death of His Son, much more, having been reconciled, we shall be saved by His life. And not only that, but we also rejoice in God through our Lord Jesus Christ, through whom we have now received the reconciliation. Rom 5:6-11 (emphasis mine).

King David gives us some reason to use our humility/mystery file in Psalm 139

> For You formed my inward parts;
>
> You covered me in my mother's womb.
>
> I will praise You, for I am fearfully and wonderfully made;
>
> Marvelous are Your works,
>
> And that my soul knows very well.
>
> My frame was not hidden from You,
>
> When I was made in secret,
>
> And skillfully wrought in the lowest parts of the earth.
>
> Your eyes saw my substance, being yet unformed.
>
> And in Your book they all were written,
>
> The days fashioned for me,
>
> When as yet there were none of them. Ps 139:13-16

Although I don't understand what I read about my lifespan in Psalm 139, I accept it all as God's Word, and file what I don't understand in my mystery file.

> All Scripture is given by inspiration of God, and is profitable for doctrine, for reproof, for correction, for instruction in righteousness, that the man of God may be complete, thoroughly equipped for every good work. 2Tim 3:16-17

Chapter 21

It's A Matter of Life and Death

Let's take a trip through the 66 books, beginning to end, and see if the Bible presents the evidence we need to understand the reality we want and need to know. Our first reference is the familiar Eden encounter.

> Then the Lord God took the man and put him in the Garden of Eden to tend and keep it. And the Lord God commanded the man, saying, "Of every tree of the garden you may freely eat; but of the tree of the knowledge of good and evil you shall not eat, for in the day that you eat of it you shall surely die." Gen 2:15-17

> Then the serpent said to the woman, "You will not surely die. For God knows that in the day you eat of it your eyes will be opened, and you will be like God, knowing good and evil." Gen 3:4-5

Conclusion, Adam lived 930 years and then he died! Somebody lied, it was the Father of Lies, the first of his many. Cities, towns and churches around the world contain the evidence of the lie, in their cemeteries. But let's continue. The next stop is the book some scholars think is Moses's first, that's Job.

> For I know that my Redeemer lives,
>
> And He shall stand at last on the earth;
>
> And after my skin is destroyed, this I know,

That in my flesh I shall see God,

Whom I shall see for myself,

And my eyes shall behold, and not another. Job 19:25-27

This declaration is the first and most complete description of life and death as we know it. When Martha met Jesus, just before He resurrected Lazarus, Jesus asked her if she thought she would see her brother again.

> Jesus said to her, "Your brother will rise again." Martha said to Him, "I know that he will rise again in the resurrection at the last day." Jesus said to her, "I am the resurrection and the life. He who believes in Me, though he may die, he shall live. And whoever lives and believes in Me shall never die. Do you believe this?" John 11:23-26

It's likely that Martha was a student of the available scriptures, specifically the writings of Moses, from which she learned about the resurrection that Job described. Curiously enough, the primary religious teachers of the day were called Pharisees and Sadducees. The Pharisees believed in the Greek idea of an immortal soul, and the Sadducees believed that death was extinction. Martha knew the scriptures and believed neither of the "experts".

In one of His discussions with the Pharisees, Jesus used their belief in an immortal soul as a way to teach an important truth, without them getting hung up on a doctrinal point that would distract them from what He wanted them to hear. We know the parable as "the rich man and Lazarus." As previously noted, a parable is a fictitious story designed to convey a specific lesson.

The Genesis account of the creation of Adam is the factual account that helps us with our understanding of the nature of life and death.

> And the Lord God formed man of the dust of the ground, and breathed into his nostrils the breath of life; and man became a living soul. Gen 2:7

Adam's creation is described as a two-step process. God formed his body, something like a potter might make a vase. At that stage, we might call Adam a "dead soul." That changed when God imparted His life to Adam by "breathing into Adam's nostrils the breath of life", thereby sharing His life with His human son, and making him a "living soul". Adam's creation differed from the prior, word creation of plants and animals. Adam was a hands on project. God's sharing His life with His son Adam was a very intimate occasion, hence the title "FATHER."

We now know how Adam became a living soul, and lived 930 years before he died. God told the truth; Satan lied-and Adam died. So, what happened to Adam when he died? The same thing that has happened to all his descendants.

> His spirit departs, he returns to his earth;
>
> In that very day his plans perish. Ps 146:4

> For what happens to the sons of men also happens to animals; one thing befalls them: as one dies, so dies the other. Surely, they all have one breath; man has no advantage over animals, for all is vanity. All go to one place: all are from the dust, and all return to dust. Who knows the spirit of the sons of men, which goes upward, and the spirit of the animal, which goes down to the earth? So I perceived that nothing is better than that a man should rejoice in his own works, for that is his heritage. For who can bring him to see what will happen after him? Eccl 3:19-22

Jesus declined to heal Lazarus as requested, to the puzzlement of the disciples, in order to demonstrate His power of life over death. The resurrection of the decayed body of His friend is the evidence on which the gospel rests.

> Our friend Lazarus sleeps, but I go that I may wake him up."

> Then His disciples said, "Lord, if he sleeps he will get well." However, Jesus spoke of his death, but they thought that He was speaking about taking rest in sleep.

Then Jesus said to them plainly, "Lazarus is dead. And I am glad for your sakes that I was not there, that you may believe. Nevertheless let us go to him." John 11:11-15

Paul sums up the Biblical teaching in his letter to the Corinthians when he says,

Now this I say, brethren, that flesh and blood cannot inherit the kingdom of God; nor does corruption inherit incorruption. Behold, I tell you a mystery: We shall not all sleep, but we shall all be changed: in a moment, in the twinkling of an eye, at the last trumpet. For the trumpet will sound, and the dead will be raised incorruptible, and we shall be changed. For this corruptible must put on incorruption, and this mortal must on immortality. So when this corruptible has put on incorruption, and this mortal has put on immortality, then shall be brought to pass the saying that is written: "Death is swallowed up in victory." 1 Cor 15:5054

Then I saw a great white throne and Him who sat on it, from whose face the earth and the heaven fled away. And there was found no place for them. And I saw the dead, small and great, standing before God, and books were opened. And another book was opened, which is the **Book of Life.** And the dead were judged according to their works, by the things which were written in the books. The sea gave up the dead who were in it, and Death and Hades delivered up the dead who were in them. And they were judged, each one according to his works. Then Death and Hades were cast into the lake of fire. This is the second death. And anyone not found written in the Book of Life was cast into the lake of fire. Rev 20:11-15

The prophet Daniel is one of the few Bible celebrities that have been awarded the accolade, "greatly beloved" by God. That's not to say God doesn't love us all. Daniel's last recorded conversation with an angel ends with the following instruction,

"But you, go your way till the end; for **you shall rest,** and **will arise to your inheritance at the end of the days."** Dan 12:13 (emphasis mine).

Job, Daniel and Paul died, looking forward to the resurrection that we find described in both the Old and New testaments. When the Lord wanted Moses with Him in Heaven, He did it by resurrection.

> Michael the archangel, in contending with the devil, when he disputed about the body of Moses, dared not bring against him a reviling accusation, but said, "The Lord rebuke you!" Jude 9-10

Paul continues his teaching on the subject,

> But now Christ is risen from the dead, and has become the first fruits of those who have fallen asleep. For since by man came death, by Man also came the resurrection of the dead.

> For as in Adam all die, even so in Christ all shall be made alive. 1 Cor 15:20-22

There are just three exceptions to the resurrection references given above. The Bible describes the exceptions as the translation of Enoch, Elijah and the people alive at the second coming of Christ. These individuals gained their eternal status without dying.

The church that Christ established understands that worshiping their Creator God is appropriate. Satan also has worshipers, in the Pagan religions of our world. Pagans believe the lie Satan told in the Garden, that man can be his own god. He can worship whatever or whomever he wishes and he will not die. The Bible consistently teaches that **we are souls, not that we have a soul.** The immortal soul idea was advanced by Satan in the garden and has no basis in scripture.

> For the living know that they will die;

> But the dead know nothing,

> And they have no more reward,

> For the memory of them is forgotten.

> Also their love, their hatred, and their envy have now perished;

> Nevermore will they have a share

In anything done under the sun. Eccl 9:5-6

In his book, THE FIRE THAT CONSUMES, Edward Fudge says, "The immortality of the soul was a principal doctrine of the Greek philosopher, Plato, who was born about the time the last Old Testament book was being written." Fudge then quotes Robert L. Wilken who writes, "The fathers' modified the notion of the immortality of the soul as it was understood with the Greek philosophical tradition. Yet, in its main lines, **they adopted the idea,** adapting it where necessary to the requirements of Christian faith and they gave it a prominent place in Christian piety."

Chapter 22

Creations-Old and New

"In the beginning, God." That's the answer to the "big bang" theory. What follows in Genesis is a description of our world being prepared for our first parents, Adam and Eve. Following Moses description of each of the days of creation, we find God's assessment of His work as "very good". Then, God put Adam and Eve in a garden and the record says He came to walk with them in the evening, as if they were good friends.

We don't know how long that happy state lasted, but in time the relationship ended when they fell for Satan's lie and became estranged from their Creator friend. Their choice left God with a dilemma. Looking down the stream of time, He could see the likes of Hitler, Stalin, and Putin, Satan's servants, living for ever in pursuit of the vilest evil. There must be limits. So, God modified His creation in consideration of the potential evil of rebellious mankind.

To the woman He said:

"I will greatly multiply your sorrow and your conception;

In pain you shall bring forth children;

Your desire shall be for your husband,

And he shall rule over you."

Then to Adam He said, "Because you have heeded the voice of your wife, and have eaten from the tree of which I commanded you, saying, 'You shall not eat of it':

"Cursed is the ground for your sake;

In toil you shall eat of it

All the days of your life.

Both thorns and thistles it shall bring forth for you,

And you shall eat the herb of the field.

In the sweat of your face you shall eat bread

Till you return to the ground,

For out of it you were taken;

For dust you are,

And to dust you shall return." Gen 3:16-19

These changes in the original creation required changes in its DNA, in order to limit the catastrophic consequences of the violation of the earth's "law of being." (Shades of the two tone, straight eight Pontiac). These necessary changes were a modification of the original perfection, in order to moderate the effects of evil. They were, in effect, a second creation.

About 1500 years from Eden, we find that mankind had broken through the barriers to evil that God had established after the fall.

Then the Lord saw that the wickedness of man was great in the earth, and that every intent of the thoughts of his heart was only evil continually. And the Lord was sorry that He had made man on the earth, and He was grieved in His heart. So the Lord said, "I will destroy man whom I have created from the face of the earth, both man and beast, creeping thing and birds of the air, for I am sorry that I have made them." But Noah found grace in the eyes of the Lord. Gen 6:5-8

> And the Lord said, "My Spirit shall not strive with man forever, for he is indeed flesh; yet his days shall be one hundred and twenty years." Gen 6:8

Another DNA modification was in order to limit the amount of evil that any one man could do.

This is the genealogy of Noah. Noah was a just man, perfect in his generations. Noah walked with God. And Noah begot three sons: Shem, Ham, and Japheth.

> The earth also was corrupt before God, and the earth was filled with violence. So God looked upon the earth, and indeed it was corrupt; for all flesh had corrupted their way on the earth. And God said to Noah, "The end of all flesh has come before Me, for the earth is filled with violence through them; and behold, I will destroy them with the earth. Make yourself an ark of gopher wood; make rooms in the ark, and cover it inside and outside with pitch. Gen 6:9-15

> Now the flood was on the earth forty days. The waters increased and lifted up the ark, and it rose high above the earth. The waters prevailed and greatly increased on the earth, and the ark moved about on the surface of the waters. And the waters prevailed exceedingly on the earth, and all the high hills under the whole heaven were covered. The waters prevailed fifteen cubits upward, and the mountains were covered. And all flesh died that moved on the earth: birds and cattle and beasts and every creeping thing that creeps on the earth, and every man. All in whose nostrils was the breath of the spirit of life, all that was on the dry land, died. So He destroyed all living things which were on the face of the ground: both man and cattle, creeping thing and bird of the air. They were destroyed from the earth. Only Noah and those who were with him in the ark remained alive. And the waters prevailed on the earth one hundred and fifty days. Gen 7:17-24

The evidence for Noah's world-wide flood is in the sedimentary rocks worldwide, where we find the burial grounds of animals, washed together by a violent hydrologic event and preserved by being buried in mud before they decayed. Many of the skeletons are incomplete, having been ripped apart by some violent force. In the Siberian Artic, huge musk oxen have been excavated with their flesh intact and food in their mouths; animals weighing tons were frozen almost instantly, while eating. Earth's history is undeniably catastrophic, and the findings are compatible with a third creation-like event, the flood of Noah. Jesus acknowledged it as reality. The current model of earth's geologic history is that the original earth was a single continent, Pangea. That model is entirely compatible with Genesis and Noah's flood.

> In the six hundredth year of Noah's life, in the second month, the seventeenth day of the month, on that day all the **fountains of the great deep were broken up, and the windows of heaven were opened.** And the rain was on the earth forty days and forty nights. Gen 7:11-12 (emphasis mine)

This text is compatible with the reconfiguration of the earth, resulting in multiple continents, mountain uplifts and plate tectonics. In His effort to further restrain evil after the flood, we find God reducing the human lifespan, adding animals to the human diet, instituting capital punishment and separating the land masses with oceans, to limit the spread of evil.

The flood teaches us that God has set limits on evil. He gives time for us to make our choice, but the time limit is real. David puts it like this,

> God is a just judge,
>
> And God is angry with the wicked every day.
>
> If he does not turn back,
>
> He will sharpen His sword;
>
> He bends His bow and makes it ready.
>
> He also prepares for Himself instruments of death;
>
> He makes His arrows into fiery shafts. Ps 7:11-13

Peter adds a balancing dimension to David's description of God's justice.

> But, beloved, do not forget this one thing, that with the Lord one day is as a thousand years, and a thousand years as one day. The Lord is not slack concerning His promise, as some count slackness, but is longsuffering toward us, not willing that any should perish but that all should come to repentance. 2 Peter 3:8-9

Scripture has several examples of God's forbearance preceding His justice. Noah built the ark and preached for 120 years before the flood. It was about 1500 years from Eden until the flood. Some people have said that Noah is the worst evangelist in history having preached 120 years to make 8 converts, and all of them were family members. I submit that the 120 years reflected God's forbearance rather than the time necessary to build the ark.

When God called Abram to go out from his home in Haran to the "promised land", He warned him that his possession would be delayed because the Amorites, who occupied the land, **had not yet filled their cup of iniquity.**

> Now when the sun was going down, a deep sleep fell upon Abram; and behold, horror and great darkness fell upon him. Then He said to Abram: "Know certainly that your descendants will be strangers in a land that is not theirs, and will serve them, and they will afflict them four hundred years. And also the nation whom they serve I will judge; afterward they shall come out with great possessions. Now as for you, you shall go to your fathers in peace; you shall be buried at a good old age. But in the fourth generation they shall return here, for **the iniquity of the Amorites is not yet complete."** Gen 15:12-16 (emphasis mine)

Another name for the cup of iniquity is probation; a time to repent. One of most profound examples is that of God's mercy as seen in the book of Jonah. God sent Jonah to the wicked city to preach about its imminent destruction.

Then word came to the king of Nineveh; and he arose from his throne and laid aside his robe, covered himself with sackcloth and sat in ashes. And he caused it to be proclaimed and published throughout Nineveh by the decree of the king and his nobles, saying, Let neither man nor beast, herd nor flock, taste anything; do not let them eat, or drink water. But let man and beast be covered with sackcloth, and cry mightily to God; yes, let everyone turn from his evil way and from the violence that is in his hands. Who can tell if God will turn and relent, and turn away from His fierce anger, so that we may not perish? Then God saw their works that they turned from their evil way; and God relented from the disaster that He had said He would bring upon them, and He did not do it. Jonah 3:6-10

Jesus's disciples looked for His return during their life-time, as have billions of His followers for over two thousand years. Some of His followers have set dates and constructed scenarios that might precede His coming. All this effort ignores Jesus's answer to the disciples when they asked when He would return, "Only the Father knows, He said."

Mankind, however, may now be able to bring about the end of the present age, with a nuclear war that would likely sterilize earth's population. It is a very scary fact of history that mankind has never invented a weapon of war that hasn't been used. The U.S. did use two small nuclear weapons to end the Second World War, when we were the only nation that possessed the bomb. Widespread use of the many nuclear weapons in existence today, would undoubtedly sterilize mankind. How close are we humans to filling our cup of iniquity?

Ronald L. McCartney, M.S., M.D.

Chapter 23

The Big Fold

The Bible gives us the history of God's missionary people, Abraham and his descendants, who were commissioned to be His witnesses to the whole world. The Redeemer was to be born of Abraham's family. The rejection of Jesus by His own people activated God's plan B, and resulted in the birth of the Christian Church, which inherited the commission, "Go into all the world" and make converts to the Creator, Father God.

Today, we have three groups of people who claim Abraham as the father of their faith: Christians, Jews and Islam. Those three faiths only account for a fraction of the world's population. How about all those other people? Jesus answered the question as follows,

> I am the good shepherd; and I know My sheep, and am known by My own. As the Father knows Me, even so I know the Father; and I lay down My life for the sheep. **And other sheep I have which are not of this fold**; them also I must bring, and they will hear My voice; and there will be one flock and one shepherd. John 10:14-16 (emphasis mine)

On another occasion, Jesus taught His disciples a little ecumenism.

Now John answered and said, "Master, we saw someone casting out demons in Your name, and we forbade him because he does not follow with us." But Jesus said to him, **"Do not forbid him,** for he who is not against us is on our side." Luke 9:49-50 (emphasis mine)

In one of the question and answer sessions between Jesus and His elite religious critics, they quizzed Him about the Law of Moses, which they accused Him of breaking.

And behold, a certain lawyer stood up and tested Him, saying, "Teacher, what shall I do to inherit eternal life?" He said to him, "What is written in the law? What is your reading of it?" So he answered and said, "'You shall love the Lord your God with all your heart, with all your soul, with all your strength, and with all your mind,' and 'your neighbor as yourself.'" Luke 10:25-27

But he, wanting to justify himself, said to Jesus, "And who is my neighbor?"

Then Jesus answered and said: "A certain man went down from Jerusalem to Jericho, and fell among thieves, who stripped him of his clothing, wounded him, and departed, leaving him half dead. Now by chance a certain priest came down that road. And when he saw him, he passed by on the other side. Likewise a Levite, when he arrived at the place, came and looked, and passed by on the other side. But a certain Samaritan, as he journeyed, came where he was. And when he saw him, he had compassion. So he went to him and bandaged his wounds, pouring on oil and wine; and he set him on his own animal, brought him to an inn, and took care of him. On the next day, when he departed, he took out two denarii, gave them to the innkeeper, and said to him, 'Take care of him; and whatever more you spend, when I come again, I will repay you.' So which of these three do you think was neighbor to him who fell among the thieves?" And he said, "He who showed mercy on him." Then Jesus said to him, "Go and do likewise." Luke 10:29-37 .

One of the great mysteries of the Bible is the account of Abraham's rescue of his nephew from the heathen kings.

> Now when Abram heard that his brother was taken captive, he armed his three hundred and eighteen trained servants who were born in his own house, and went in pursuit as far as Dan. He divided his forces against them by night, and he and his servants attacked them and pursued them as far as Hobah, which is north of Damascus. So he brought back all the goods, and also brought back his brother Lot and his goods, as well as the women and the people. And the king of Sodom went out to meet him at the Valley of Shaveh (that is, the King's Valley), after his return from the defeat of Chedorlaomer and the kings who were with him. Then **Melchizedek king of Salem** brought out bread and wine; he was the **priest of God Most High**. And he blessed him and said:

> "Blessed be Abram of God Most High,

> Possessor of heaven and earth;

> And blessed be God Most High,

> Who has delivered your enemies into your hand."

> And he gave him a tithe of all. Gen 14:14-20 (emphasis mine)

The identity of this "priest of God" other than his being a priest and king, is unknown. He apparently ruled the ancient city, Salem, the precursor of Jerusalem. Abraham thought enough of him to pay him a tithe. Just one of Jesus's "other sheep"? Paul identifies some of those other sheep like this,

> Do not lie to one another, since you have put off the old man with his deeds, and have put on the new man who is renewed in knowledge according to the image of Him who created him, where there is neither Greek nor Jew, circumcised nor uncircumcised, barbarian, Scythian, slave nor free, but **Christ is all and in all.** Col 3:9-11 (emphasis mine)

Chapter 24

Out of this World

During the twentieth century, the development of large terrestrial telescopes expanded our view of the universe beyond our wildest imagination. In 1990, the Hubble telescope was placed into an earth orbit of about 350 miles. It became operational in December of 1993. The pictures of our universe that it provided were nothing short of astounding, and expanded the dimensions of space far beyond our previous concept.

Edwin Hubble, for whom the space scope was named, had discovered galaxies beyond our own, while working at the Mount Wilson Observatory in California. Many of what had been considered stars by previous astronomers, have turned out to be galaxies, measuring thousands of light years distant from our planet, and some of them larger than our own Milky Way. Thanks to the space scope we now have an even more extensive view of "deep space" not previously possible.

The discovery of an extensive, infinite universe outside our home galaxy has opened significant questions about the age of our universe. The Hubble findings challenge the conventional young earth creationists' understanding of Genesis chapter one. In a recent DVD, produced by five highly qualified Creation scientists, they address what has been called the Starlight Travel Dilemma." They propose a number of theories which attempt to explain how the universe, including deep space, could be only about six thousand years old, in keeping with their interpretation of Genesis chapter one. Other observers say that the deep star light we see has been on its way for much longer than a few thousand years.

In my attempt to follow their theories I was reminded of Occam's rule and I re-read the Genesis account in the light of some relevant, related Biblical references. Having been a creationist of long standing, and a believer in the inspired Word, I felt compelled to try to harmonize the evidence, like Thomas Aquinas had.

I remembered that Moses wrote the book of Genesis sometime on the forty year journey with recently freed slaves, who had no county of their own. They were to be the nation promised by God, to Abraham from which the Deliverer promised in Eden would come. The book of Genesis was the charter for that new nation. The book documents the preparation of our pre-existent planet for life and it gave the descendants of Abraham a unique identity, history and purpose.

The word "heavens" is the point of this discussion. What did Moses mean by his use of the word heaven? Help is available! A previous quotation from Paul noted his "trip" to the **third Heaven** which he had no words to describe. It was obviously the home of the Deity. Moses defines his use of the word "heaven" by describing the parts of our galaxy we use for times, seasons and navigation. All visible to the **unaided eye.** When Absalom's rebellion failed, he fled and was caught by his long hair in the branches of a tree. He was said to have been suspended between **"Heaven and Earth".**

So we have **three heavens** described in scripture. The context is determinative. When Moses says "In the beginning", the most logical meaning is the beginning of our habitable planet. The idea that the eternal God had to have made deep space at the same time He enlivened our planet, defies the context and purpose of Genesis. I don't see a deep space star-light problem. My consideration of the "star light problem" brought to mind Dr. Smith.

In 1955 the first year medical class in human physiology met with Dr. Maynard Smith for the first time. He announced that he would begin with a quiz! Grumble, grumble the class. By a show of hands we were to answer the following, "What is America's favorite sport?" He said, "Raise your hand if you think it is baseball." About a third of the class raised a hand. "You're wrong," he snapped. "How many of you think its football?" he queried. About half the class raised a hand. "Not even close," he retorted. Pause, shuffle, shuffle, walking back and forth. "Ok, ok, how many of you think its sex?" Big sigh across the class, that's got to be it, as they raised their hands and voices, after all this guy is a physiologist!

After some order had returned, Dr. Smith announced, with unmistakable authority, "You're all wrong, and I've noticed that some of you aren't thinking! That's a very bad thing in my class." Now that he had everybody's attention, he answered his own question, "It's jumping," he said, "jumping to conclusions."

The lecture that followed was on the methodology of empherical science.

He said that the best explanation of any observed phenomenon is

reached when enough quality, harmonious data is available to form a hypothesis, which can be evaluated by qualified observers and given theory status if verified (falsified). He endorsed Occam's razor, favoring the simplest possible explanation to the problem under consideration.

Testing (falsifying) a theory by multiple objective investigators establishes or disproves a theory. When the available data is too limited for a convincing conclusion, the true scientist must resist the temptation to jump to conclusions based on his/her desired outcome, philosophy or religion. Although everyone is entitled to his/her opinion, when their ideas are not verified by pure empherical science or by credible scholarship, based on historical data, they should not represent their **opinion** as settled science. That standard of honesty should especially apply when the unproven or controversial idea is being advance by persons in positions of power or authority like teachers, preachers and scientists.

An example is in order. In France, the Pasteur family was blessed with a son. The year was 1822. Louis grew into one of the most productive scientists of any age, in spite of it being the nineteenth century, which is on record as one of the most turbulent times in the sciences, as well as in religion and philosophy.

A popular theory of the day was that life had its origin in death. The evidence, it was said, is everywhere. Dead bodies of humans and animals are covered with maggots soon after death, flies emerge from rotting fruit, mice come out of piles of decaying burlap bags and leaches are found in the mud along the bands of lakes the streams.

A British scientist, John Neeham, devised an experiment to affirm the popular belief that life can originate from lifeless matter. One such idea was that sunlight acting on pond water creates the many single cell "animules" observed through the microscope. He designed an experiment in which he boiled pond water to kill the life forms he had demonstrated, then put a beaker of that sterilized pond water on a window sill, where it was exposed to the sun and air in the open container. A few weeks later, the beaker was the swimming pool for innumerable one celled "animules." WHALA, the scientific evidence for spontaneous generation of life. Who needs God?

Pasteur observed that life comes/came from life, parents and God respectively, and he found Neeham's experiment flawed. So, he redesigned the experiment to conform to true empirical principles, eliminating all but one variable. After boiling the pond water to kill the life forms, he sealed the sterile water in a glass vessel and placed it on the window exposed to sunlight. He had eliminated exposure of the sterile water to the open air, thereby eliminating the second variable. It was just sterile pond water and sun now. Weeks later, there was no growth. Now opening the vessel, he exposed it to the air and the life forms that circulate in the wind soon produced florid growth. Two scientists, one real, the other a practitioner of agenda science.

Pasteur repeated his experiment many times in various venues and at various altitudes. He demonstrated that there are more life forms circulating in the wind at lower altitudes than in the higher elevations. He published his findings and conclusions in 1859, the same year that Charles Darwin published his ORIGIN OF SPECIES, which curiously enough proposed a theory based on the idea that Pasteur had disproved!

Chapter 25

We Get to Choose

One might argue, on the basis of the Old Testament Biblical record, that God's plan to redeem His creation failed. That was likely Satan's conclusion on Good Friday, when he succeeded in convincing the Jews to reject their own Messiah, by crucifixion. The record from Easter morning, thru the gospels to John's Revelation, says God wins. In fact, it was Satan and his followers who lost from the start, and they will continue their loosing ways until their "cup" is full, God's kingdom is complete and the judgement ends evil, forever. God's patience is why we have evil in our world. The world's "cup" must be about full!

In the conversation Jesus had with Nicodemus, He makes the distinction between an eternal perspective and the temporal one. Jesus makes the point that mankind, is by birth, oriented to earth and time, but that God offers us, as He did with Adam and Eve, life with an eternal perspective. Paul explains the born again analogy like this,

> Therefore, if anyone is in Christ, he is a **new creation**; old things have passed away; behold, all things have become new. 2 Cor 3-11 (emphasis mine)

Thomas Aquinas understood that the visible (science) and the invisible (religion) worlds were both made by the Creator God, and must therefore be in agreement. Further, said Aquinas, if there appears to be a disagreement between the temporal and spiritual explanations, the spiritual explanation takes president. William Paley, the father of Intelligent Design, wrote about finding a watch on the garden path and concluding that someone unknown to him had designed and constructed it. His unimpeachable logic is the reasonable basis for the Intelligent Design explanation of our universe.

I find it fascinating that Richard Dawkins, world famous evolutionist, agrees with King David, Paul, Aquinas and Paley that the universe looks as if it had been designed. But then he jumps the constraints of logic and concludes that it wasn't. Dr. Smith was right again!

The apostle Paul told the Romans that Dawkins is, in Paul's words, a fool,

> For the wrath of God is revealed from heaven against all ungodliness and unrighteousness of men, who suppress the truth in unrighteousness, because what may be known of God is manifest in them, for God has shown it to them. For since the creation of the world His invisible attributes are clearly seen, being understood by the things that are made, even His eternal power and Godhead, so that they are without excuse, because, although they knew God, they did not glorify Him as God, nor were thankful, but became futile in their thoughts, and their foolish hearts were darkened. Professing to be wise, they became fools, and changed the glory of the incorruptible God into an image made like corruptible man — and birds and four-footed animals and creeping things. Rom 1:18-23

Dawkins' conclusion is one of the two available explanations for the universe. Adam and Eve had to choose between the same two options, as have all their descendants. We can embrace the Greek idea of impersonal, autonomous natural forces which have created us and our universe or we can acknowledge the logical conclusion of our observations that scream design and dependence.

The first option focuses on the temporal, and has no concept or consciousness of the eternal. Epicurus said it best, "Eat, drink and make merry, for tomorrow we die"; and death is annihilation. The Greek god(s) were created by men, and endowed with human character flaws, so that the Pagan could conveniently blame his own failings on the gods.

The Christian option helps us face the reality of our dependence on the Father in heaven for our creation and sustenance. As God's children we learn to grow up into His full design for us, embrace the law of our being, and live our lives in the anticipation of eternal fellowship with Him and our brothers and sisters in the "great cloud of witnesses". We have thereby been "born again" into the eternal dimension.

Chapter 26

Our Eternal Witness

Jesus's disciples were brought up in the Jewish Messianic tradition that originated in Eden, some four thousand years prior. That tradition had changed in the interval, however, from the original goal of delivery from eternal death, to a temporal goal of national political freedom. The Jews were looking for a deliverer who would free them from the Romans or other temporal overlords.

Jesus's challenge was to change the disciples' perspective from earth to heaven, death to life and temporal to eternal. He had to change their minds and elevate their vision, so as to validate their witness and empower their efforts to spread the good news.

> Then He said to them, "These are the words which I spoke to you while I was still with you, that all things must be fulfilled which were written in the Law of Moses and the Prophets and the Psalms concerning Me." And He opened their understanding, that they might comprehend the Scriptures. Then He said to them, "Thus it is written, and thus it was necessary for the Christ to suffer and to rise from the dead the third day, and that repentance and remission of sins should be preached in His name to all nations, beginning at Jerusalem. And **you are witnesses of these things.** Behold, I send the Promise of My Father upon you; but tarry in the city of Jerusalem until you are endued with power from on high." Luke 24: 44-49 (emphasis mine)

"Go into all the world and preach the gospel to every creature. He who believes and is baptized will be saved; but he who does not believe will be condemned. And these signs will follow those who believe: In My name they will cast out demons; they will speak with new tongues; they will take up serpents; and if they drink anything deadly, it will by no means hurt them; they will lay hands on the sick, and they will recover." Mark 16:15-18 (emphasis mine)

The first recorded evidence of the disciples having gotten the big picture is in Peter's witness to the Jewish leaders.

"Men and brethren, let me speak freely to you of the patriarch David that he is both dead and buried, and his tomb is with us to this day. Therefore, being a prophet, and knowing that God had sworn with an oath to him that of the fruit of his body, according to the flesh, He would raise up the Christ to sit on his throne, the, foreseeing this, spoke concerning the resurrection of the Christ, that His soul was not left in Hades, nor did His flesh see corruption. **This Jesus God has raised up, of which we are all witnesses.** Therefore being exalted to the right hand of God, and having received from the Father the promise of the Holy Spirit, He poured out this which you now see and hear. Acts 2:29-33 (emphasis mine)

In addition to witnessing that Jesus was the long promised Messiah, the message that the followers of Jesus preached was of His second coming, to eliminate evil and to install righteousness. The disciples hoped that the second coming would happen in their day, just like Jesus's followers have wished for the last two thousand years. Peter explained the apparent delay in the following quotations:

For we did not follow cunningly devised fables when we made known to you the power and coming of our Lord Jesus Christ, but were eyewitnesses of His majesty. For He received from God the Father honor and glory when such a voice came to Him from the Excellent Glory: "This is My beloved Son, in whom I am well pleased." And we heard this voice which came from heaven when we were with Him on the holy mountain. 2 Peter 1:16-18

But, beloved, do not forget this one thing, that with the Lord one day is as a thousand years, and a thousand years as one day. The Lord is not slack concerning His promise, as some count slackness, but is longsuffering toward us, **not willing that any should perish** but that all should come to repentance. 2 Peter 3:8-9 (emphasis mine)

In the words of an old gospel song, "When He cometh when He cometh to **make up His jewels,** all His jewels, precious jewels, His Loved and His own…" The words remind us that Lucifer was created the "Covering Cherub". He ranked fourth in the heavenly courts, the prime minister of the Deity, a jewel in the crown. By his insurrection he **deprived God of His principle witness** and one third of the angelic hosts. Might we, Christ's redeemed, be the eternal witnesses to replace that loss? Maybe "making up His jewels" explains what we think is a delay?

Behold what manner of love the Father has bestowed on us, that we should be called children of God! Therefore the world does not know us, because it did not know Him. Beloved, **now we are children of God;** and it has not yet been revealed what we shall be, but we know that when He is revealed, **we shall be like Him,** for we shall see Him as He is. And everyone who has this hope in Him purifies himself, just as He is pure. 1 John 2:1-3 (emphasis mine)

A Definition Of Christian Faith

"Faith" is a word we use to describe a relationship with God as with a Person well known. The better we know Him, the better the relationship may be.

Faith implies an attitude toward God of love, trust, and deep admiration. It means having enough confidence in God, based upon the more than adequate evidence revealed, to be willing to believe whatever He says, to accept whatever He offers, and to do whatever He wishes—without reservation—for the rest of eternity. Anyone who has such faith is perfectly safe to save. This is why faith is the only requirement for heaven.

From the Director, Division of Religion, Loma Linda University